Recycling of Polyethylene Terephthalate Bottles

Plastics Design Library

PLASTICS DESIGN LIBRARY (PDL) PDL HANDBOOK SERIES
Series Editor: Sina Ebnesajjad, PhD (sina@FluoroConsultants.com) President, FluoroConsultants Group, LLC Chadds Ford, PA, USA
http://www.FluoroConsultants.com

The **PDL Handbook Series** is aimed at a wide range of engineers and other professionals working in the plastics industry, and related sectors using plastics and adhesives.

PDL is a series of data books, reference works and practical guides covering plastics engineering, applications, processing, and manufacturing, and applied aspects of polymer science, elastomers and adhesives.

Recent titles in the series

Introduction to Plastics Engineering, Shrivastava (ISBN: 9780323395007)
The Effect of Sterilization Methods on Plastics and Elastomers, McKeen (ISBN: 9780128145111)
Polymeric Foams Structure-Property-Performance, Obi (ISBN: 9781455777556)
Technology and Applications of Polymers Derived from Biomass, Ashter (ISBN: 9780323511155)
Fluoropolymer Applications in the Chemical Processing Industries, 2e, Ebnesajjad & Khaladkar (ISBN: 9780323447164)
Reactive Polymers, 3e, Fink (ISBN: 9780128145098)
Service Life Prediction of Polymers and Plastics Exposed to Outdoor Weathering, White, White & Pickett, (ISBN: 9780323497763)
Polylactide Foams, Nofar & Park (ISBN: 9780128139912)
Designing Successful Products with Plastics, Maclean-Blevins (ISBN: 9780323445016)
Waste Management of Marine Plastics Debris, Niaounakis, (ISBN: 9780323443548)
Film Properties of Plastics and Elastomers, 4e, McKeen, (ISBN: 9780128132920)
Anticorrosive Rubber Lining, Chandrasekaran (ISBN: 9780323443715)
Shape-Memory Polymer Device Design Safranski & Griffis, (ISBN: 9780323377973)
A Guide to the Manufacture, Performance, and Potential of Plastics in Agriculture, Orzolek, (ISBN: 9780081021705)
Plastics in Medical Devices for Cardiovascular Applications, Padsalgikar, (ISBN: 9780323358859)
Industrial Applications of Renewable Plastics, Biron (ISBN: 9780323480659)
Permeability Properties of Plastics and Elastomers, 4e, McKeen, (ISBN: 9780323508599)
Expanded PTFE Applications Handbook, Ebnesajjad (ISBN: 9781437778557)
Applied Plastics Engineering Handbook, 2e, Kutz (ISBN: 9780323390408)
Modification of Polymer Properties, Jasso-Gastinel & Kenny (ISBN: 9780323443531)
The Science and Technology of Flexible Packaging, Morris (ISBN: 9780323242738)
Stretch Blow Molding, 3e, Brandau (ISBN: 9780323461771)
Chemical Resistance of Engineering Thermoplastics, Baur, Ruhrberg & Woishnis (ISBN: 9780323473576)
Chemical Resistance of Commodity Thermoplastics, Baur, Ruhrberg & Woishnis (ISBN: 9780323473583)
Color Trends and Selection for Product Design, Becker, (ISBN: 9780323393959)
Fluoroelastomers Handbook, 2e, Drobny (ISBN: 9780323394802)
Introduction to Bioplastics Engineering, Ashter (ISBN: 9780323393966)
Multilayer Flexible Packaging, 2e, Wagner, Jr. (ISBN: 9780323371001)
Fatigue and Tribological Properties of Plastics and Elastomers, 3e, McKeen (ISBN: 9780323442015)
Emerging Trends in Medical Plastic Engineering and Manufacturing, Schönberger & Hoffstetter (ISBN: 9780323370233)
Manufacturing and Novel Applications of Multilayer Polymer Films, Langhe & Ponting (ISBN: 9780323371254)

To submit a new book proposal for the series, or place an order, please contact Edward Payne, Acquisitions Editor at edward.payne@elsevier.com

Recycling of Polyethylene Terephthalate Bottles

Edited by

Sabu Thomas

International and Interuniversity Centre for Nano Science and Nanotechnology and School of Chemical Sciences, Mahatma Gandhi University, Kerala, India

Ajay Rane

Durban University of Technology, Composite Research Group, Durban, South Africa

Krishnan Kanny

Durban University of Technology, Composite Research Group, Durban, South Africa

Abitha V.K.

School of Chemical Sciences, Mahatma Gandhi University, Kerala, India

Martin George Thomas

Department of Polymer Science and Rubber Technology, Cochin University of Science and Technology, Cochin, Kerala, India

ELSEVIER

William Andrew
Applied Science Publishers

William Andrew is an imprint of Elsevier
The Boulevard, Langford Lane, Kidlington, Oxford, OX5 1GB, United Kingdom
50 Hampshire Street, 5th Floor, Cambridge, MA 02139, United States

Notices

Knowledge and best practice in this field are constantly changing. As new research and experience broaden our understanding, changes in research methods, professional practices, or medical treatment may become necessary.

Practitioners and researchers must always rely on their own experience and knowledge in evaluating and using any information, methods, compounds, or experiments described herein. In using such information or methods they should be mindful of their own safety and the safety of others, including parties for whom they have a professional responsibility.

To the fullest extent of the law, neither the Publisher nor the authors, contributors, or editors, assume any liability for any injury and/or damage to persons or property as a matter of products liability, negligence or otherwise, or from any use or operation of any methods, products, instructions, or ideas contained in the material herein.

British Library Cataloguing-in-Publication Data
A catalogue record for this book is available from the British Library

Library of Congress Cataloging-in-Publication Data
A catalog record for this book is available from the Library of Congress

ISBN: 978-0-12-811361-5

Publisher: Matthew Deans
Acquisition Editor: Edward Payne
Editorial Project Manager: Mariana L. Kuhl
Production Project Manager: Kamesh Ramajogi
Cover Designer: Christian J. Bilbow

Typeset by MPS Limited, Chennai, India

Contents

List of Contributors

A.R. Ajitha International and Inter University Centre for Nanoscience and Nanotechnology, Mahatma Gandhi University, Kottayam, Kerala, India

M.K. Aswathi International and Inter University Centre for Nanoscience and Nanotechnology, Mahatma Gandhi University, Kottayam, Kerala, India

Subhendu Bhandari Maharashtra Institute of Technology, Aurangabad, Maharashtra, India

Karan R. Chavan Department of Chemical Engineering, Institute of Chemical Technology, Mumbai, India

Cintil Jose Chirayil Department of Chemistry, Newman College, Thodupuzha, Kerala, India

Ayan Dey Department of Plastic and Polymer Engineering, Maharashtra Institute of Technology, Aurangabad, Maharashtra, India

K.G. Gopalakrishna Center for Incubation, Innovation, Research and Consultancy, Jyothy Institute of Technology, Bangalore, India

Prashant Gupta Maharashtra Institute of Technology, Aurangabad, Maharashtra, India

Myungwan Han Department of Chemical Engineering & Applied Chemistry, Chungnam National University, Daejeon, Korea

Krishnan Kanny Composite Research Group, Department of Mechanical Engineering, Durban University of Technology, Durban, South Africa

Suranjana Mandal Department of Plastic and Polymer Engineering, Maharashtra Institute of Technology, Aurangabad, Maharashtra, India

P. Manju Department of Plastics Technology, Central Institute of Plastics Engineering and Technology, Chennai, Tamil Nadu, India

K.V. Marathe Department of Chemical Engineering, Institute of Chemical Technology, Mumbai, India

Raghvendra Kumar Mishra International and Inter University Centre for Nanoscience and Nanotechnology, Mahatma Gandhi University, Kottayam, Kerala, India

Pranav Nakhate Department of Chemical Engineering, Institute of Chemical Technology, Mumbai, India

Rabindra Kumar Padhan Department of Civil and Environmental Engineering, Hong Kong Polytechnic University, Hong Kong, P.R. China

Deepak Pant Department of Environmental Sciences, Central University of Himachal Pradesh, Dharamshala, Himachal Pradesh, India

Ajay Rane Composite Research Group, Department of Mechanical Engineering, Durban University of Technology, Durban, South Africa

Narendra Reddy Center for Incubation, Innovation, Research and Consultancy, Jyothy Institute of Technology, Bangalore, India

Anvita Sheel Department of Environmental Sciences, Central University of Himachal Pradesh, Dharamshala, Himachal Pradesh, India

Anand Sreeram Department of Civil and Environmental Engineering, Hong Kong Polytechnic University, Hong Kong, P.R. China

Sabu Thomas International and Interuniversity Centre for Nanoscience and Nanotechnology and School of Chemical Sciences, Mahatma Gandhi University, Kottayam, Kerala, India

1 PET Chemistry

Suranjana Mandal and Ayan Dey
Department of Plastic and Polymer Engineering, Maharashtra Institute of Technology, Aurangabad, Maharashtra, India

1.1 Introduction

Polyethylene terephthalate or poly(ethylene terephthalate), abbreviated as PET, it is one of the most abundantly available semicrystalline thermoplastic resin and it is the only fiber and film forming saturated polymer with commercial importance [1].

This is the commercially popular polymer used in fibers for clothing, liquid, and food container. In addition PET has good mechanical property (such as tensile and tear strength), heat sealability, excellent barrier property against oxygen, carbon dioxide, anhydride, aroma compound, etc. [2]. This polymeric resin is also used in the production of thermoformed structure as well as used in combination of glass fiber as composite.

W.H. Carothers, an American Chemist and inventor of Nylon first synthesized thermoplastic polyesters from adipic acid and ethylene glycol (EG), in 1929. In 1941, high molecular weight polyesters based on terephthalic acid (TPA) was synthesized, which became a product of industrial interest. J.R. Whinfield and J.T. Dickson of Calico Printers Association in England, invented poly(ethylene terephthalate) fiber (Terylene, Dacron, Lavsan) and also film (Melinex, Mylar) and moulding material which becomes promising candidate for producing blow moulded bottles.

In the 1950s, Du Pont and Imperial Chemical Industries, Ltd., began manufacture of fiber melt−spun synthetic polyester products. Yet it was not until the mid-1960s that thermoplastic polyesters were employed as construction materials. These were fast-crystallizing grades of PET containing additives that gave uniform and controlled morphology.

A short time later, copolyesters were developed with enhanced properties. PET containers formed by stretch blow molding were introduced in the United States by Du Pont in the mid-1970s. Since that time, the market for PET has grown at 10% per year, with initial growth spawned

Recycling of Polyethylene Terephthalate Bottles.
DOI: https://doi.org/10.1016/B978-0-12-811361-5.00001-8

by direct replacement of glass and metal, as well as the introduction of new products such as heat-set containers.

Today, PET is used to produce a variety of products, including filament and staple fiber, film, tire cord, technical yarns, and packaging resins. PET may be produced from EG with either purified terephthalic acid (PTA) or dimethyl terephthalate (DMT) in presence of a metal ion (catalyst). Till 1960, DMT was preferred as feedstock for PET manufactures because of the fact that ester could generally be made in more pure form than acid. Later, Amoco developed high purity terephthalic acid process, which becomes the preferred feedstock for the production of PET. Advantage of using highly PTA eliminates the need of recycling methanol, rapid esterification of the prepolymer and initial step of esterification process (when DMT is in use as feedstock).

1.2 Physical and Chemical Properties

PET, a type of thermoplastic engineering polyester possesses excellent mechanical, electrical, and thermal properties. Moreover it has very good chemical resistance, transparency, dimensional stability, and low moisture absorption. The important properties are enlisted below [3,4]:

1. Transparent and light weight
2. Extremely low moisture diffusion through the polymer matrix
3. Exceptional dimensional stability
4. Excellent electrical property
5. Excellent chemical resistance
6. Good environmental stress crack resistance
7. Very good resistance to thermal aging
8. Very low creep at wide temperature range
9. Very good color stability
10. Excellent wear resistance
11. Low flavor adsorption
12. Better aesthetic appearance (compared to other polymers)
13. Economic availability and easy process ability
14. Excellent recyclability

Higher ordered regularities in chemical and geometric architecture of the polymer result in crystalizability of the PET matrix. Either it is semicrystalline or amorphous in nature. PET with high crystallinity possesses higher glass transition temperature compared to amorphous PET along with mechanical properties like higher modulus, toughness, stiffness, tensile strength, hardness, and more solvent resistance. The impact strength of the material becomes inferior with the increase in crystallinity [5].

Fig. 1.1 shows the impact of the ester group concentration on the melting point of different linear polyester. The structures of homologous polyesters are shown in Table 1.1. For the linear aromatic

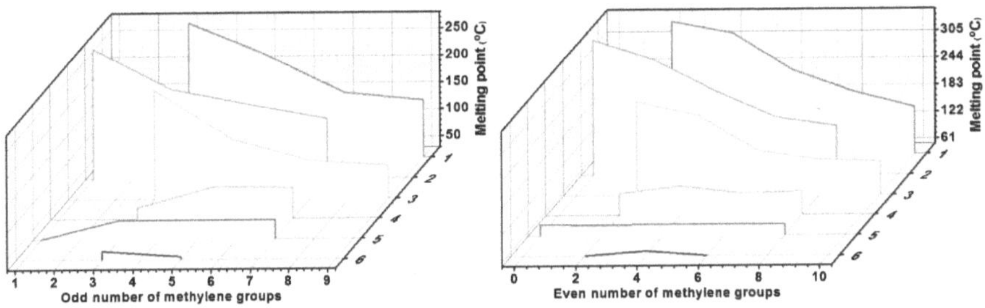

Figure 1.1 Melting points (T_m) of some homologous polyesters [5].

Table 1.1 Homologous Polyesters With Repeat Unit Structures [5]

Series	Structure of Repeat Units
1. Poly(alkylene biphenyl-4, 4'-dicarboxylate)	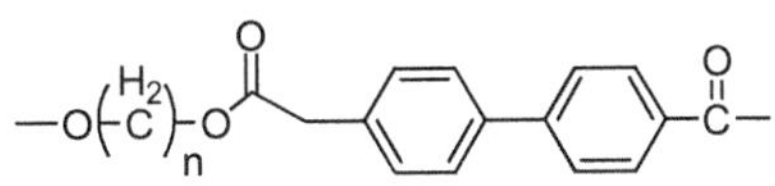
2. Poly(p-phenylene alkandioate)	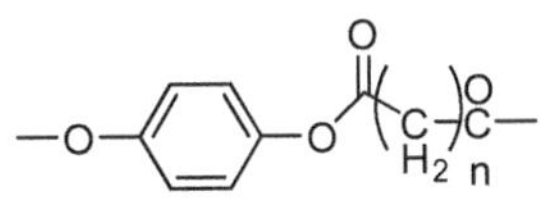
3. Poly(alkylene terephthalate)	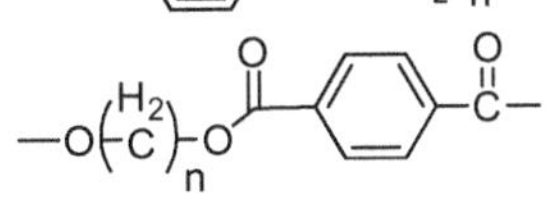
4. Poly(alkylene succinate)	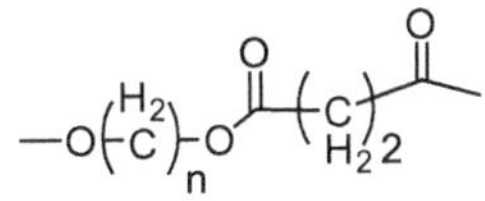
5. Poly(decamethylene alkane dioate)	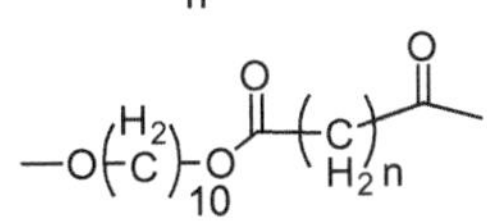
6. Poly(alkylene adipate)	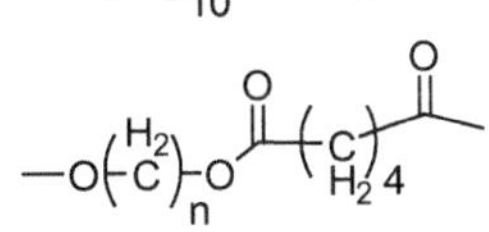

polyesters, melting point decreases with the decrease in the concentration of ester groups. However, the variation of melting points with the increasing number of methylene group in between two ester linkages present in the backbone is of conspicuous nature.

To resolve the mystery, the data of melting points of polyesters available in the literature have been plotted separately for even and odd numbered of carbon atoms. The variation of the melting point has been found to be same for both of them with increasing number of methylene groups in between two ester linkages. The melting point has been found to decrease exponentially and attains a plateau when number of methylene groups reaches to six. However, it can also be said that decrease in concentration of ester group causes a slight increase in the melting point.

The polyesters are illustrated in Table 1.2. With the increase in methylene groups in the repeating unit, the polymer behaves like linear

Table 1.2 Reagents for PET Synthesis [6,7]

Type	Molecule	Molecular Structure	Influencing Property
Diols	Ethylene glycol		Fiber grade PET formation
	Diethylene glycol	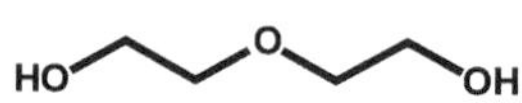	Improves dye ability of fiber grades and lowers melting temperature
	Bisphenol A	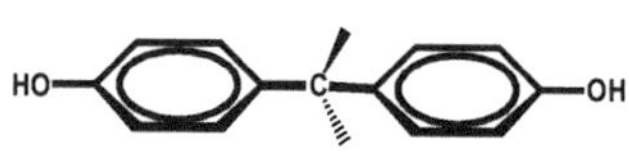	Improves melt stiffness
	EG end group		Decrease in hydroxyl value which improves thermal stability
	DEG end group	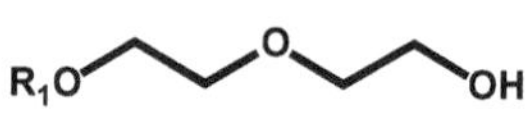	Decrease in hydroxyl value which improves thermal stability and flexibility
	bDEG repeat unit	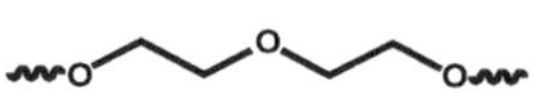	Improvement in mechanical property
	Vinyl end group		End group functionalization of PET

Table 1.2 Reagents for PET Synthesis [6,7]—cont'd

Type	Molecule	Molecular Structure	Influencing Property
Diacids	Terephthalic acid		Formation of carboxyl end groups and acetaldehyde results in yellowing of the polymer and reduction in the hydrolytic and thermal stability
	Isophthaleic acid		Stress cracking resistance and melting temperature
	2,6-Naphthalene dicarboxylic acid	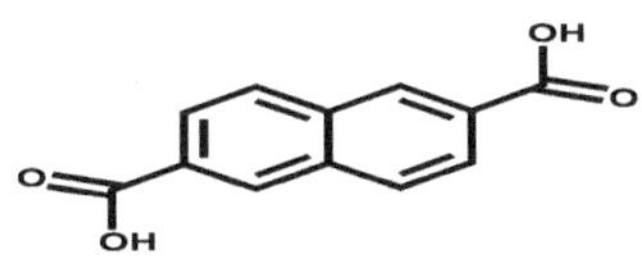	Improves mechanical properties and reduces gas permeability
	TPA repeat unit		Improves the hydrolytic and thermal stability as acid value reduces

polyethylene (polymethylene). Thus the melting points for five polyesters among the classes available are seen to converge towards that of the polymethylene. In the case of the sixth class, the poly(alkylene adipates) behaves differently. It is interestingly observed that the melting point for polymers with off numbered methylene groups in the aliphatic moiety is significantly lower than the melting point for the polymers composed of even number of methylene groups. But, this is seen to be valid only for those polymers obtained by condensation and rearrangement polymerization methods.

The density of a crystalline PET is 1.455 g/cm^3 whereas the density of amorphous PET is 1.333 g/cm^3 [8]. However, PET crystallizes slowly in absence of nucleating agents and plasticizers which creates hindrance during processing through injection moulding. In case of films crystallinity is induced by mechanical operations (e.g., biaxially oriented PET films). In addition, the annealing is another process which is carried out by heating the polymer film from 180 to 220°C for several seconds to allow amorphous chain relaxation. This may results in recrystallization, and crystal growth to occur in PET matrix. PET is chemically inert in nature so it requires variety of surface modifier. The drawing of the film after the application of coating helps to achieve very thin layers of the coating materials.

1.3 Synthesis and its Chemistry

Monomers used to synthesize PET are mainly TPA and EG. Various dicarboxylic acids and aliphatic or aromatic diols may be incorporated in order to modify the polymer characteristics. Most important reagents for synthesizing PET are shown in Table 1.2. Synthesis of PET involves four stages: (1) transesterification/direct esterification, (2) prepolymerization, (3) melt condensation, and (4) solid-state polycondensation. The formation of PET chains can occur namely by two main reactions: (1) esterification of carboxyl groups with hydroxyl end groups, and (2) transesterification (transalcoholysis) of glycolesters end terminated with hydroxyl groups. The reverse reactions of the aforesaid reactions are hydrolysis and glycolysis, respectively.

1. In the first step, bis(hydroxyethyl)terephthalate (BHET) and some of its linear oligomers is synthesized by transesterification of DMT or direct esterification of terephthalic acid with EG (Fig. 1.2). However, the transesterification procedure was proved as a most efficient process to be used in order to develop PET fiber. DMT is synthesized by esterification reaction between TPA and EG at a temperature of 240–260°C and a pressure of about 300–500 kPa [7].
2. The direct esterification is also used to synthesize BHET. The mechanism is shown in Fig. 1.3. Direct esterification of acid and glycol has been evolved as one of the commercial manufacturing method. In direct esterification, TPA and EG react to produce the DMT intermediate. Advantages of this process are as follows:
   a. Easy purification of ester compared to acid.
   b. Improved catalytic action.
   c. Easy elimination of methanol which is obtained as by-product during transesterification.

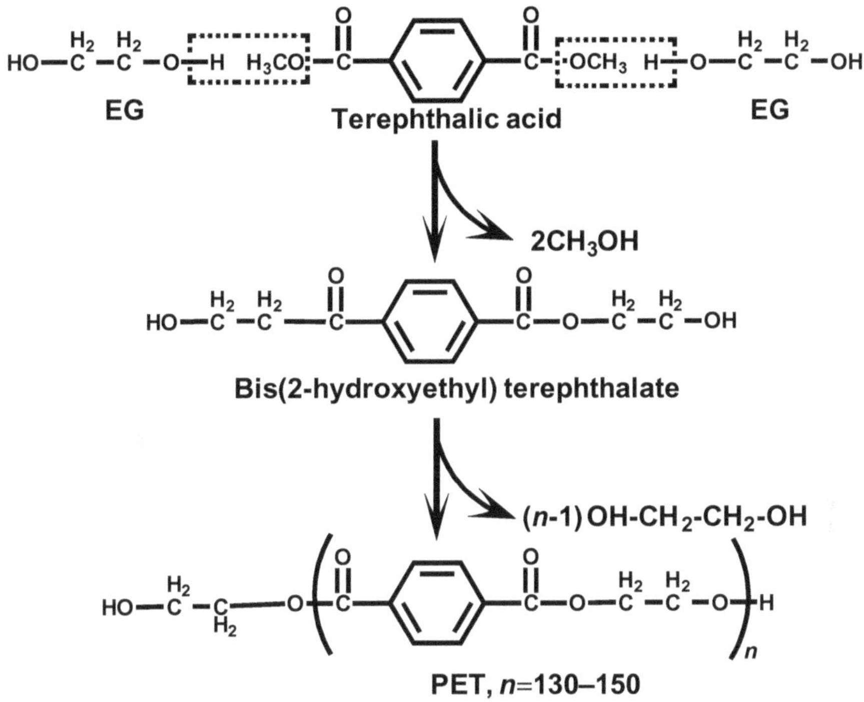

Figure 1.2 Synthesis of bis(2-hydroxyethyl)terephthalate and polymerization of the same to PET.

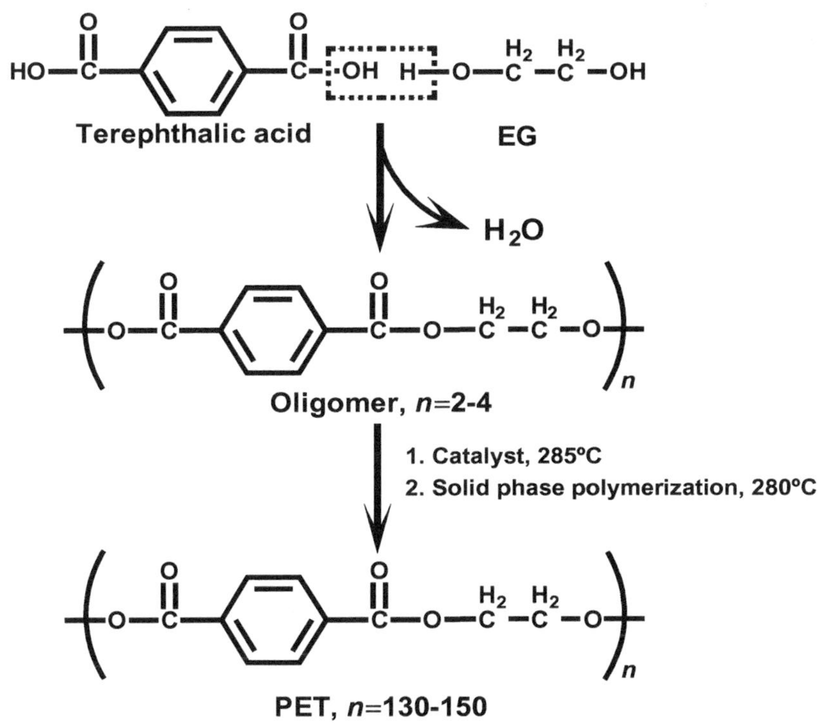

Figure 1.3 Manufacture of PET by direct esterification reaction [9].

Generally, PET resins synthesized by this process contain more diethylene glycol, which is by-product obtained due to an intermolecular ether-forming reaction among β-hydroxyethyl ester end groups. Oriented films made by these resins have inferior mechanical rigidity and comparatively low melting points, poor thermo-oxidative resistance as well as inferior UV stability [10].

3. The second step involves the conversion of BHET (and some oligomers) emerging from the first stage to a prepolymer up to a degree of polymerization (DP) of ~30 and the viscosity of ~5.0 Pa S [10]. Here, the EG undergoes reaction with DMT at 150–2200°C at 100 kPa [11].
4. In the third stage, the product is then allowed to polymerize to a DP about 100.
5. The final stages of polycondensation reaction are carried out in a single reactor equipped with conventional helical screw. The process is known as solid-state polycondensation reaction which leads to a product of DP about 150. The catalyst used for polycondensation for transesterification reaction at this stage is antimony trioxide (Sb_2O_3) or antimony triacetate. However, antimony pentaoxide and germanium dioxide have also been used for much reaction [10]. 0.5%–2% of a diphenyl ester, preferably diphenyl terephthalate is introduced into the reaction mixture as chain extender during transesterification reaction. It is not only accelerating the production rate but also improves the color of finished product.

The ester interchange reaction is carried out in presence of catalyst, usually transition metal salts like manganese, cobalt or zinc or alkaline earth metal, such as calcium. At this stage, the PET is a viscous molten liquid. This is extruded and water quenched to form a transparent amorphous material. Another form of PET is also manufactured using dimethyl ester of terephthalic acid (DMT). The required high molecular weight PET is manufactured by solid-state polymerization (SSP) at relatively lower temperatures. This form of PET is free from all volatile impurities like acetaldehyde, free glycols, and water.

1.4 Continuous Melt Phase Polymerization

Here, polymerization reaction is a type of step growth reaction occurs through condensation of condensable groups. The high molecular weight is essential for good mechanical properties like stiffness and creep resistance simultaneously providing sufficient flexibility to resist bursting and fractured under pressure. Purification of the polymer is very difficult after polymerization. So, purity of the starting materials is important.

Vacuum distillation processes easily purify EG while terephthalic acid is purified by repeated crystallization. Such high purity and high molecular weight materials are needed for food packaging applications. Catalysts are used at extremely low concentrations for promoting the reactions and ensure practical economics. The most common catalyst is antimony trioxide. However salts of titanium, germanium, cobalt, manganese, magnesium, and zinc are also used [11]. Moreover, PET can be synthesized using terephthaloyl chloride with EG. In spite of high reaction rate and yield such reaction is not used due to its high cost and hazardous characteristic of the reagent used [12].

The steps required to produce high molecular weight polymers:

1. EG remains unreacted during alcoholysis must be completely removed to ensure good conversion rate.
2. The temperature of the reaction medium must be controlled during the second stage of the reaction (i.e., propagation step) to eliminate the possibility of thermal chain scission and thermal decomposition of end group (it results in formation of acetaldehyde as by-product). Weight–average molecular weight of oriented PET films obtained in this way is around 35,000 g/mol.

1.5 Catalyst Chemistry and Mechanism

Acid catalyzed esterification technique is used to synthesize PET. Order of the reaction is 3 (2 with respect to acid and 1 with respect to alcohol). The application of metal catalysts in polyesterification has been described by Fradet and Maréchal [13]. The catalyst is used here performs as both reactant as well as catalyst. Tetrabutoxytitanium is a very efficient catalyst, compared to zinc acetate or diantimony trioxides are not for polyesterification [14]. The catalyst used for polycondensation for transesterification reaction at this stage is antimony trioxide (Sb_2O_3) or antimony triacetate. In addition, the polycondensation catalysts also include germanium and titanium compounds. Yamada reported the effect of potassium titanium oxyoxalate [PTO; $K_2TiO(C_2H_4)_2.2H_2O$)] as a polycondensation catalyst. In addition, same article also reported the higher efficiency of the catalyst with respect to antimony catalyst. PTO is observed to have higher acceleration ability on the main reaction and less influence on side reactions. Moreover, it also results in higher reduction of carboxylic acid group concentration in the PET obtained compared to other catalyst system [15]. However, antimony pentaoxide and germanium dioxide have also been used for such reaction [10]. Toshiro Yamada evidences the diantimony trioxide acts as positive catalyst in production of PET [15]. Antimony or titanium compounds are promising candidate as catalysis for the polycondensation reaction by ligand-exchange reactions. Weingart [16] demonstrated that the titanium ions are able to form coordinate covalent bond to the oxygen atom of fast-exchanging alkoxy ligands. The polycondensation reaction is favored by the formation of chelates. Siling and Laricheva reviewed extensively on titanium catalysts for esterification as well as transesterification [17]. The mechanism of titanium based catalyst system for PET synthesis is described in Fig. 1.4.

However, titanium catalyst results in formation of titanium vinylates which results in yellowing of PET. Weingart and coworkers found that the degree of yellowing depends on the catalyst concentration [18]. Finelli et al. describes the effect of titanium tetrabutoxide on the properties of PET. The article also explained the influence of the catalyst

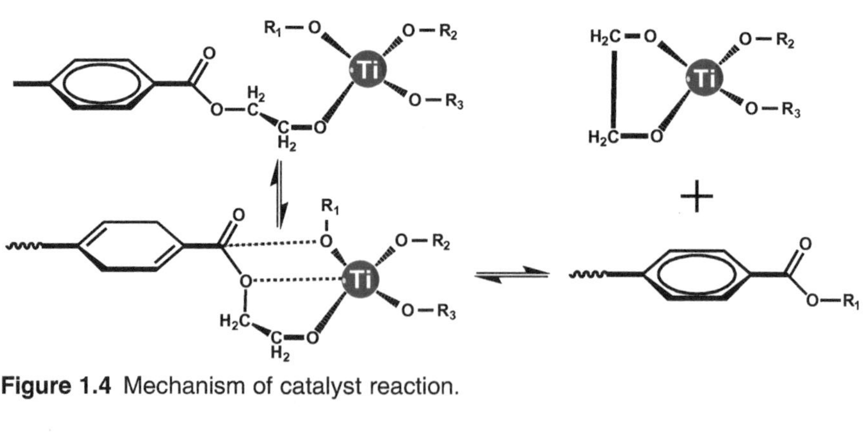

Figure 1.4 Mechanism of catalyst reaction.

concentration on the molecular weight and melting temperature of PET [19]. The catalyst contains a significant amount of alkoxides which are responsible for the almost similar concentration of the acid value or carboxylic end group concentration and DEG or hydroxyl value observed for PET fiber obtained. The article explains the mechanism of esterification reaction. Some side reaction also occurred during the reaction which leads to evolution of diethylene glycol. Sb_2O_3 only accelerates the principle reaction only yielding the esters of terephthalic acid [15]. The article also evidences that increasing concentration of the antimony oxide results in decrease in carboxylic end group concentration. However, higher concentration of catalyst also results in agglomeration which shortens the polymer chain formed. So, it is desirable to keep it at minimum during production of PET. A mechanism of titanium based catalyst system for PET synthesis is described in Fig. 1.4.

Although most of the manufacturers use antimony oxide to catalyze transesterification reaction, antimony acetates are also used as transesterification catalyst [20]. The efficiency of the antimony compound is enhanced in presence of esters but it is independent on the acid group concentration. The antimony glycolate formed during the reaction is less reactive than the β-hydroxy ester. Thus elimination of EG obtained during the reaction is performed to resist the poisoning of the catalytic activity of the aforesaid catalyst. Antimony compound is not introduced in the reaction vessel when BHET is produced from the reaction EG and DMT. However, Sb-based catalysts can be introduced when high concentration of the acid group is present in the reaction aliquot and overall process can be catalyzed spontaneously [20].

Youk et al. reported the use of antimony oxide to polymerize ethylene terephthalate cyclic oligomers. The antimony catalyst results in high molecular weight PET when polymerization reaction occurs at 293°C. Only 0.5 mol.% is sufficient to produce PET possessing number average molecular weight of 22,000 [21]. In absence of the catalyst, only 4800 will be the value of number average molecular weight achievable under the same reaction condition (293°C). The highest molecular weight has been reported to obtain at 0.3 mol.% diantimony trioxide concentration. However, the resultant molecular weight of PET was reported to be independent of the catalyst [21]. Duh et al. describes the mechanism of catalytic action of diantimony trioxide for solid-state esterification reaction to obtain PET. The investigation of Shah et al. [22] evidences that antimony possesses higher activity compared to titanium and tin towards the polycondensation of bis(2-hydroxyethyl) terephthalate at molten condition. Such catalyst possesses minimum tendency to catalyze degradation reactions. This is the most suitable polycondensation catalyst among the inorganic catalysts like Ti, Sn, Sb, Mn, Zn, and Pb. Mechanism of catalytic action is described in Fig. 1.5. If Sb_2O_3 is added directly to the reaction vessel, it is rapidly converted to glycolate in presence of free glycols. EG is obtained as side product, which promotes the formation of antimony glycolate. In the subsequent step, the end group coupled with the catalyst can react with unreacted end group to obtain a diester linkage. This results in regeneration of the glycolate. This glycolate continue to participate in the similar reaction [23]. The mechanism is proposed by Parshall and Ittel [24].

One of the most salient features of antimony catalyst is the low catalytic activity in thermal degradation. Oxidation state of Sb is +3 which comprises of one lone pair. Sb^{3+} is amphoteric in nature which

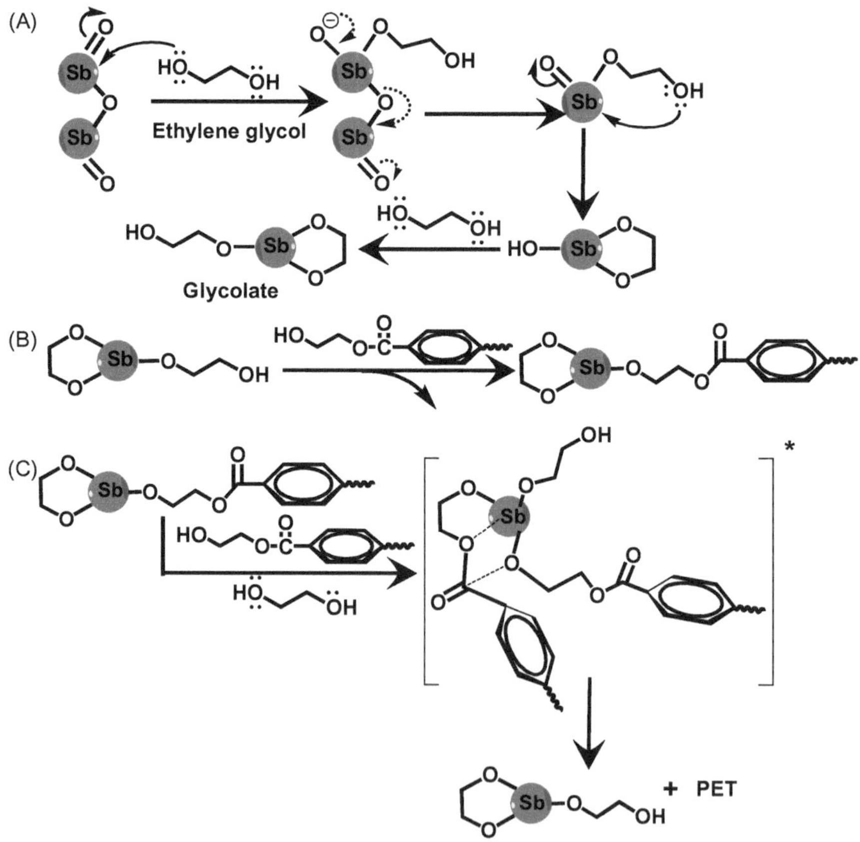

Figure 1.5 Mechanism of catalytic activity of antimony oxide.

suppresses interaction of the metal ion with carbonyl oxygen. A coordinate bond can be formed among the alcoholic oxygen of ester groups and the antimony ion. As hydroxyl group is more nucleophilic than alcoholic oxygen of ester group, which may react preferentially with the antimony compounds of weak electrophilic characteristics. However, such coordination complex formation is favored by building stable chelate system. The lone pair of electrons for antimony also causes interference in the stereochemistry of its complexes due to its repulsive effect on the ligands. Therefore, the reacting groups are compelled to come closer to each other allowing the reaction among them. This results in switching the bonds among the groups in chelate which will result in polyester formation. El-Toufaili et al. describes the step by step mechanism during the PET synthesis [25].

According to industry reports, PET has a worldwide production exceeding 30 million metric tons per annum (MTA). Most of it is homopolymer, primarily used for the manufacture of staple and filament fibers. Approximately one-third of all PET produced today is "solid state" using a SSP process. The SSP process is accomplished by heating the polymer in solid-chip form at a temperature above the T_g of the polymer (80°C) but below its melting point (~250°C). The SSP process typically is conducted in nitrogen or helium gaseous (inert gas) environment to prevent degradation of the polymer at the elevated temperature. In solid stating, similar reactions take place to those in the melt phase polymerization process. The molecular weight of PET is increased by condensation polymerization (chain growth). SSP was first introduced for the production of industrial technical fibers such as tire cord, which require higher molecular weights to produce the desired mechanical strength and tenacity. SSP, by definition, is the further processing of a polymer in the "solid state" (more correctly, semi-solid, or "glassy," state above the glass transition temperature) to increase the molecular weight to much higher levels than can be accomplished easily in the melt phase. The main objective of the SSP process is to increase the polymer's molecular weight, density, and viscosity, thus improving the physical and mechanical properties for select applications [17].

1.6 Chain Extender

0.5%–2% of a diphenyl ester, preferably diphenyl terephthalate is introduced into the reaction mixture as chain extender during

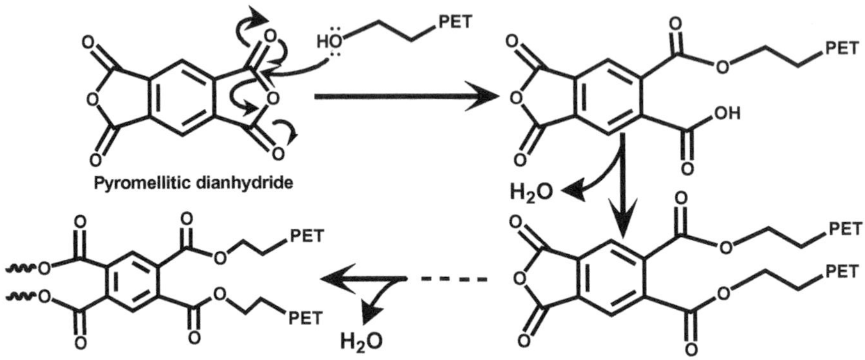

Figure 1.6 Step by step mechanism of reaction of chain extender.

transesterification reaction. It is not only accelerates the production rate but also improves the color of finished product. Moreover, use of chain extender increases melt viscosity as molecular weight of the polymer increases. It also improves the mechanical rigidity of the polymer matrix. The molecules having functionality greater than two can be used as chain extender. Awaja et al. reported use of pyromellitic dianhydride as chain extender for PET [26]. The step by step mechanism is described by Fig. 1.6. 1.1,1.3-Phenylene bis(2-oxazoline-2) and isophthaloyl biscaprolactamate are also reported by Loontjens et al. as chain extender for PET [27]. It is recommended by Loontjens et al. to use two types of chain extenders to increase in molecular weight of PET. Synergistic approach of use both chain extenders at the same time results an enormous increase in melt strength of the PET [27]. Raffa et al. reported the use of multifunctional epoxy or isocyanate additives as chain extender [28]. The di-isocyanate exhibits higher reactivity than the di-epoxide with the terminals of PET. A dramatic increase in melt viscosity is reported in presence of aforesaid polyfunctional chain extenders compared to the PET synthesized using difunctional ones at similar molar concentration of overall functional group. However, variation of the molecular weight and melt viscosity are observed to reach a plateau when the concentration of the additive used reaches a threshold value. According to Raffa et al., the value is "5 mmol chain extender/ 100 g PET" [28]. In addition, chain extender is also used to reduce acid value of the PET. The said additive is used to react with carboxylic acid end group which results in sharp reduction of the acid value. The reduction of acid value not only suppresses the color generation in the product during service life but also improves the oxidative resistance of the PET polymer.

1.7 Structure–Property Relationship of PET

PET contains ester linked backbone which can go for hydrolytic (under acidic or alkaline condition) and thermal cleavage. This restricts the use of the material for high temperature applications [29]. However, linear saturated and unsaturated polyester resins are used as hot melt adhesives. In addition, PET has also found wide applicability as fibers and films.

Polyester is extensively used to extend the service life of shoe polyester–amide copolymers has also gains immerse interest in automobile industries [30]. The different applications of the polyesters require different structure and functional groups present on the polymer backbone, the nature of end group, molecular weight of repeat unit, controls the physical and chemical properties of the polymer. Thermal and thermomechanical aspects depend on flexibility of the polymer chains.

Chain flexibility is of the two types:

1. Kinetic chain flexibility and
2. Thermodynamic flexibility

The flexibility of the polymer chain can be controlled by introducing some spacer molecule either as a part of polymer backbone or as a part of blend with the polymer. Kinetic chain flexibility of a polymer is defined by an extent of maximum uncoiling under the action of the field applied (i.e., tension, shear or thermal application). The polymer chains having low potential rotation barriers possess great kinetic flexibility.

Kinetic flexibility increases with

1. Increase in polymer hydrocarbon chain length
2. Polymer whose chains have the segments like ether, ester and –NH–(C = O)–O–– which are separated by long chain methylene group
3. Polymers having polar substituents arranged sparsely along the chain
4. Polymer having short chain branching (with low degree of branching)

Polymer with higher concentration of polar group exhibits higher kinetic rigidity. The polymer backbone containing ring structure is kinetically rigid high kinetic flexibility results in lowering glass transition temperature. However, the polymers having higher kinetic flexibility bear low thermodynamic flexibility [31].

Kinetic flexibility can be controlled by:

1. Controlling the molar mass of a polymer.
2. Lowering the length of Kuhn segment.
3. Formation of optical network.
4. Inducing dipole–dipole interaction or increasing orientation.
5. Crystallinity of the polymer.

Kinetic flexibility bears a proportional relationship with molecular mass [31]. Polymers having low molecular mass have higher chain flexibility than that for the high molecular weight polymers. The presence of chemical bonds between chains limit their mobility thus routes the kinetic rigidity of the polymer chains.

Kuhn segment (b) is another parameter which can limit the kinetic chain flexibility. Kuhn segment is defined by segmental length to, i.e., the chain length of the segment after which the continuity of the chain brakes. It depends on the Flory theta conditions (i.e., the condition of unoriented amorphous bulk r_θ).

$$b = \frac{r_\theta^2}{L}$$

Increase in Kuhn segment for polar polymers results in increase in kinetic rigidity of the polymer chain.

Increasing orientation factors may also results in reduction of kinetic flexibility. The higher degree of orientation improves the packing of density of the molecule which in turn results in high degree of interchange interaction. This interaction restricts the molecular movement inducing the chain rigidity in the polymer matrix. Similarly, crystallinity also results in reducing chain flexibility of the polymer matrix.

PET contains a rigid ring structure as well as small number of methylene group attached to it. The structure depicts the high rigidity of the polymer matrix.

However, it can be reduced by copolymerizing with other monomer (as described in Table 1.2) and by modifying the end group characteristics. Introduction of small number of methyl groups may also be used for lowering the chain rigidity. The phenomenon is described in Fig. 1.7. Controlling chain rigidity is essential parameters to control the mechanical and thermal characteristics. The specified applications of the solid materials create immerse thus in finding the ways to engineer the molecule architecture. The additives and catalyst used have

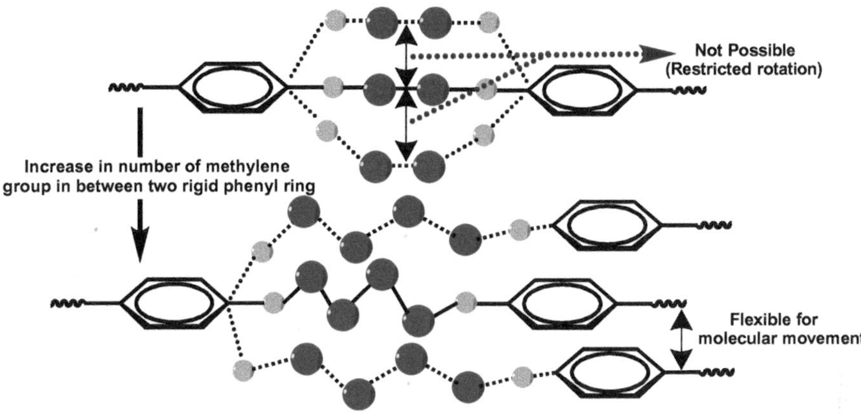

Figure 1.7 Effect of increase in number of methylene group in between two rigid phenyl ring structure.

profound impact of PET chemistry and determines the end property of the polymer matrix.

Fu et al. had attempted to draw a structure–property relationship for the PET fiber [32]. Tenacity and shrinkage are two main important parameters for PET fibers are described as a structure sensitive property. Shrinkage results mainly due to randomization of strained, oriented, noncrystalline domains. In case of rigid-amorphous fractions shrinkage can occur. Here the rigid-amorphous fraction of the polymer matrix becomes mobile. This type of transition for rigid-amorphous material is often broad and not monotonous and may occur in between T_g and T_m. The tenacity is increased with the increase in unidirectional orientation of fiber. In addition, deformation before break has been reported to increase instantaneously during orientation [32]. PET fibers can gradually reach much higher moduli without drastic changes in structure or processing mechanisms. According to Gupta et al., the PET fiber contains residual stress which gives rise to thermal shrinkage. This in turns make the polymer dimensionally unstable. Heat setting stabilizes the structure by bringing the molecular orientation in equilibrium position. PET is generally heat set at a temperature between 190 and 225°C [33]. It also affects the crystallinity of the matrix. The crystallization rate is low below T_g due to lack of mobility and above T_m due to the excess mobility of the molecules. The rate of crystallization for PET attains maxima at 180°C [33]. Heat setting at higher temperature results in increase in crystallite size. This in turn results in reduction of constrains generated in amorphous state which improves the mechanical property as well. The yield stress of the crystalline PET is proportional to the crystal size. Rastogi et al. reported about reduction in strain hardening modulus with the decrease in crystallization temperature [34] as the number of entanglements per chain in the amorphous domains decreases due to combined effect of crystalline ordering and the reeling-in effect. Increase in crystallization temperature results a decrease in crystal size as well as yield stress. Rastogi et al. evidences such increase in yield stress along with decrease in strain hardening with the increase in temperature for crystallization process of PET [35]. Li et al. reported an avenue to decrease the brittle characteristics of PET by copolymerizing with other monomer ethylene iso-phthalate [35]. The article evidences about a decrease in T_g with the increase in concentration of ethylene iso-phthalate as it disturbs the symmetry of the chain. The decrease in T_g also induces the flexibility or ductility in the polymer chain [35].

1.8 New Enzymatic Route to Synthesize Polyester

Enzymatic polymerization is a new chemical route to synthesize polymers "via biosynthetic paths by an isolated enzymes" as stated in the review of Aliki Douka et al. [36]. Benefits of such process are: (1) reaction can be carried out at low temperature, (2) nontoxic medium, (3) polymer obtained is of good quality due to high enantio, chemo and regioselectivity, (4) no side reaction, and (5) works in any types of environment (e.g., bulk, solution, suspension, biphasic organic media, etc.). Recyclability, eco-friendly, and nontoxicity made the enzymes valuable as initiator for polymerization of PET to obtain eco-friendly plastic especially important for biomedical applications. Whereas, organometallic catalysts introduces toxicity in the polymer if not completely removed. So removal of such catalytic residue is stringent for the PET which is produced through melt condensation reaction. Such postpolymerization steps are not required for enzymatic route which make the process popular in recent days. Moreover, removal of oxygen and moisture is not strictly required for this process to obtain high molecular weight PET. In addition, as the polymerization reaction temperature does not exceed 100°C for enzymatic route, the possibility of degradation and depolymerization becomes insignificant. This further improves the product life, aesthetics and resulting properties. Significant progress for producing linear polyesters of high molecular weight was made by many groups of scientists for linear aliphatic polyesters [37–40]. But work on aromatic polyesters is very limited [36]. RM and N435 were mostly used enzymes to produce aliphatic and aromatic polyester, respectively. The enzyme N435 is also effective for transterification of the diester of the diacids by diols. Novozym 435 or N435 is immobilized lipase B from Candida Antarctica. RM or Lipase RM is *Rhizomucor miehei* lipase. The polyesterification reaction using RM as enzyme was carried out at a temperature of 37°C under vacuum using diphenyl ether as solvent. Yield reported for such reactions is 85%–93% [41]. In case of using N435 as biocatalyst, similar solvent and techniques were used. Reaction rate as well as yield is also excellent for such system. However, longer chain length of the diacid possesses higher reactivity than the systems of shorter length of diacids [42].

1.9 Conclusion

This chapter reveals that not only structure of PET determines its characteristics but also the processing conditions have the impact to modify the properties of the fiber. In addition, the various additives in the form of chain extender, nucleating agents, etc. posses profound impact on the characteristics of polymers. Melting point of PET can be tuned by increasing the number of methylene groups present in between two carbonyl groups of the repeating unit. Alteration of crystallinity and crystallite size results in improvement of mechanical properties of the polymer. Disturbing symmetry of the repeating unit has been found to reduce the glass transition temperature which in turn increases the ductility in the polymer chains.

References

[1] B. Golding, Polymers and Resins: Their Chemistry and Chemical Engineering, Van Nostrand, New York, 1959.
[2] V. Siracusa, P. Rocculi, S. Romani, M. Dalla Rosa, Biodegradable polymers for food packaging: a review, Trends Food Sci. Technol. 19 (12) (2008) 634–643.
[3] J. Bjorksten, H. Tovey, B. Harker, J. Henning, Polyesters and Their Applications, Chapman and Hall, London, UK, 1956.

[4] G.N. VanderPlaats, Numerical Optimization Techniques for Engineering Design, VanderPlaats Research & Development Inc, Colorado Springs, USA, 1999.

[5] B. Demirel, A. Yaraş, H. Elçiçek, Crystallization behavior of PET materials, Balıkesir Üniversitesi Fen Bilimleri Enstitüsü Dergisi 13 (1) (2016) 26–35.

[6] J.A. Brydson, Plastics Materials, Butterworth-Heinemann, 1999, p. 713.

[7] https://www3.epa.gov/ttnchie1/ap42/ch06/final/c06s06-2.pdf.

[8] C.A. Harper, Modern Plastics Handbook: Handbook, McGraw-Hill Professional, New York, 2000.

[9] S.M. Metro, J.F. McGehee, UOP sinco solid-state polymerization process for the production of pet resin and technical fibers, in: R.A. Meyers, R.A. Meyers (Eds.), Handbook of Petrochemicals Production Processes, McGraw-Hill Prof Med/Tech., Access Engineering, 2005.

[10] K. Ravindranath, R.A. Mashelkar, Polyethylene terephthalate—I. Chemistry, thermodynamics and transport properties, Chem. Eng. Sci. 41 (9) (1986) 2197–2214.

[11] P.V. Smirnov, L.P. Repina, N.S. Bunigina, É.M. Aizenshtein, N.M. Kvasha, V.V. Kiselev, Transesterification of dimethyl terephthalate with ethylene glycol, Fibre Chem. 15 (5) (1984) 332–336.

[12] A.M. Al-Sabagh, F.Z. Yehia, G. Eshaq, A.M. Rabie, A.E. ElMetwally, Greener routes for recycling of polyethylene terephthalate, Egypt. J. Petrol. 25 (1) (2016) 53–64.

[13] A. Fradet, E. Maréchal, Kinetics and mechanisms of polyesterifications, Polymerizations and Polymer Properties., Springer, Berlin, Heidelberg, 1982, pp. 51–142.

[14] O.M.O. Habib, J. Málek, The activity of metal ions in catalytic esterification of aromatic carboxylic acids with aliphatic glycols, Collect. Czechoslov. Chem. Commun. 41 (9) (1976) 2724–2736.

[15] T. Yamada, Effect of diantimony trioxide on direct esterification between terephthalic acid and ethylene glycol, J. Appl. Polym. Sci. 37 (7) (1989) 1821–1835.

[16] F. Weingart, Titankatalysatoren bei der Herstellung von Polyethylenterephthalat – Untersuchungen zu Wirkung und Katalysemechanismus (Ph.D. thesis), University of Stuttgart, 1994.

[17] M.I. Siling, T.N. Laricheva, Titanium compounds as catalysts for esterification and transesterification, Russ. Chem. Rev. 65 (3) (1996) 279.

[18] F. Weingart, P. Hirt, H. Herlinger, Titanium catalysts in the manufacture of polyethylene terephthalate, Chem. Fibers Int. 46 (2) (1996). 96–96.

[19] L. Finelli, C. Lorenzetti, M. Messori, L. Sisti, M. Vannini, Comparison between titanium tetrabutoxide and a new commercial titanium dioxide based catalyst used for the synthesis of poly(ethylene terephthalate), J. Appl. Polym. Sci. 92 (3) (2004) 1887–1892.

[20] B. Gantillon, R. Spitz, T.F. Mckenna, The solid state postcondensation of PET, 1, Macromol. Mater. Eng. 289 (1) (2004) 88–105.

[21] J.H. Youk, R.P. Kambour, W.J. MacKnight, Polymerization of ethylene terephthalate cyclic oligomers with antimony trioxide, Macromolecules 33 (10) (2000) 3594–3599.
[22] T.H. Shah, J.I. Bhatty, G.A. Gamlen, D. Dollimore, Aspects of the chemistry of poly(ethylene terephthalate): 5. Polymerization of bis (hydroxyethyl) terephthalate by various metallic catalysts, Polymer 25 (9) (1984) 1333–1336.
[23] B. Duh, Effect of antimony catalyst on solid-state polycondensation of poly(ethylene terephthalate), Polymer 43 (11) (2002) 3147–3154.
[24] G.W. Parshall, S.D. Ittel, Homogeneous Catalysis: The Applications and Chemistry of Catalysis by Soluble Transition Metal Complexes, Wiley, New York, 1992.
[25] F.A. El-Toufaili, G. Feix, K.H. Reichert, Mechanistic investigations of antimony-catalyzed polycondensation in the synthesis of poly(ethylene terephthalate), J. Polym. Sci. A: Polym. Chem. 44 (3) (2006) 1049–1059.
[26] F. Awaja, F. Daver, E. Kosior, F. Cser, The effect of chain extension on the thermal behaviour and crystallinity of reactive extruded recycled PET, J. Therm. Anal. Calorim. 78 (3) (2004) 865–884.
[27] T. Loontjens, K. Pauwels, F. Derks, M. Neilen, C.K. Sham, M. Serne, The action of chain extenders in nylon-6, PET, and model compounds, J. Appl. Polym. Sci. 65 (9) (1997) 1813–1819.
[28] P. Raffa, M.B. Coltelli, S. Savi, S. Bianchi, V. Castelvetro, Chain extension and branching of poly(ethylene terephthalate) (PET) with di- and multifunctional epoxy or isocyanate additives: an experimental and modelling study, React. Funct. Polym. 72 (1) (2012) 50–60.
[29] I. Benedek, M.M. Feldstein (Eds.), Technology of Pressure-Sensitive Adhesives and Products., CRC Press, 2008.
[30] J.W. Gooch, Analysis and Deformulation of Polymeric Materials: Paints, Plastics, Adhesives, and Inks, Springer Science & Business Media, 1997, p. 190 (Chapter 11).
[31] A. Tager, Physical Chemistry of Polymers, MIR Publishers, Moscow, 1978.
[32] Y. Fu, B. Annis, A. Boller, Y. Jin, B. Wunderlich, Analysis of structure and properties of poly(ethylene terephthalate) fibers, Journal of Polymer Science Part B: Polymer Physics 32 (13) (1994) 2289–2306.
[33] http://nopr.niscair.res.in/handle/123456789/32299.
[34] R. Rastogi, W.P. Vellinga, S. Rastogi, C. Schick, H.E.H. Meijer, The three-phase structure and mechanical properties of poly(ethylene terephthalate), J. Polym. Sci. B: Polym. Phys. 42 (11) (2004) 2092–2106.
[35] B. Li, J. Yu, S. Lee, M. Ree, Poly(ethylene terephthalate-co-ethylene isophthalate)—relationship between physical properties and chemical structures, Eur. Polym. J. 35 (9) (1999) 1607–1610.

[36] A. Douka, S. Vouyiouka, L.M. Papaspyridi, C.D. Papaspyrides, A review on enzymatic polymerization to produce polycondensation polymers: the case of aliphatic polyesters, polyamides and polyesteramides, Prog. Polym. Sci. 79 (2018) 1–25.
[37] A. Zaks, A.M. Klibanov, Enzyme-catalyzed processes in organic solvents, Proc. Natl. Acad. Sci. 82 (1985) 3192–3196.
[38] A.M. Klibanov, Enzymes that work in organic solvents, Chemtech 16 (1986) 354–359.
[39] R. Katsarava, V. Beridze, N. Arabuli, D. Kharadze, C. Chu, C. Won, Amino-acid based bioanalogous polymers. Synthesis and study of regular poly(ester amide)s based on bis(a-aminoacid), a-ω-alkylene diesters and aliphatic dicarboxylic acids, J. Polym. Sci. A Polym. Chem. 37 (1999) 391–407.
[40] A. Chaudharry, J. Lopez, E.J. Beckman, A.J. Russell, Biocatalytic solvent-free polymerization to produce high molecular weight polyesters, Biotechnol. Prog. 13 (1997) 318–325.
[41] C.J. Morrow, Biocatalytic synthesis of polyesters using enzymes, MRS Bull. 17 (1992) 43–47.
[42] A. Mahapatro, B. Kalra, A. Kumar, R. Gross, Lipase-catalyzed polycondensations: effect of substrates and solvent on chain formation, dispersity and end-group structure, Biomacromolecules 4 (2003) 544–551.

2 Regulations on Recycling PET Bottles

K.G. Gopalakrishna and Narendra Reddy
Center for Incubation, Innovation, Research and Consultancy, Jyothy Institute of Technology, Bangalore, India

2.1 Introduction

One of the definitions for regulations is "a rule or directive made and maintained by an authority" which implies that there are some aspects that need to be restricted in order to reduce misuse or unintended consequences. Synthetic polymers and corresponding plastic products are one the most important inventions in the history of humankind. Plastic has become inevitable and an integral part of daily activities. However, the ubiquitous use and indiscriminate disposal of plastics have led to considerable human and environmental implications. Polyethylene terephthalate (PET), particularly in bottle form is one of the most largely consumed plastic products. In fact, it is difficult to find any liquid food that has or not being packaged in PET bottles. Since PET bottles are mostly intended for one-time use, most of them are immediately disposed. Hence, having good collection, recycling and reuse programs for PET bottles are important to avoid their misuse and reduce negative environmental impact.

PET bottles are mostly used for food and beverage applications. Although PET bottles are highly recyclable, it is unfortunate that the rate of PET bottle recycling is considerably low (Table 2.1). Difficulties in collection, mixture of PET bottles with other materials and modifications necessary to process recycled PET with virgin polymers contribute to the low rate of PET recycling. In addition, although a majority of PET bottles are colorless, presence of colorants is an issue for recycling. Similarly, the caps of PET bottles are generally made using polyolefin's and have to be recycled separately. In most instances, PET bottles are reprocessed for bottle to bottle use and hence, utmost care should be taken when PET bottles are recycled.

Recycling of Polyethylene Terephthalate Bottles.
DOI: https://doi.org/10.1016/B978-0-12-811361-5.00002-X

Table 2.1 Contaminants and Applications of PET Bottles Intended for Recycling

Common PET Bottle Contaminants	Intended Applications of Recycled PET Bottles
– Polyvinyl chloride (PVC) – Plastic films, bags, carrier bags – Non plastic items such as glass, paper, and silicone – Plastics such as polystyrene, HIPS, and ABS – Fine, dirt, loose labels, rubber – Adhesives	– Belts, blankets, boat hulls, business cards, beverage bottles – Car parts such as bumpers and exterior panels, carpets, egg cartons – Polyester fabric for upholstery, T-shirts, sweaters, athletic wear and shoes – Tennis ball cans – Paint brush bristles – Pillows

Due to the ecological impact of disposing PET bottles, several rules and regulations have been enacted in many developed countries. Unfortunately, regulations on recycling and reuse of PET are not prevalent in many countries even today [1]. However recycling of PET also leads to considerable contamination and health hazards and hence several regulations have to be followed for recycling of PET bottles and PET in general. The extent and type of recycling are also dependent on certain regulations, processing methods used and economical aspects. For example, mechanical recycling provides 1 kg of recycled material from 1.25 kg of raw material compared to 1 kg requiring 1.13 kg for chemical conversion (http://www.donusumpet.com/en/important-informations). However, chemical recovery and recycling is considerably expensive than mechanical processing. According to Food and Drug Administration (FDA), there are three levels of plastic scrap that are considered. First is the preconsumer or primary scrap that is generated in the production facilities. Second is the physical reprocessing where the product is treated to remove contaminants and formed in flakes or pellets for further processing. Third is chemical reprocessing wherein the plastic is depolymerized to recover the original monomer or oligomer. Some references have also given a fourth classification of recycling (quaternary recycling) where plastics are converted into energy [2]. Specific terminology is used for the type and process of collecting plastics for recycling. Curbside collection, drop-off and subscription programs have been developed for effective collection of waste plastics.

Similarly, single stream or dual stream is used to describe the presence of one or more types of plastics.

Many countries have established bottle to bottle recycling facilities. Although Germany was one of the leading countries in terms of recycling PET, developing countries such as India are recently reporting recycling rates as high as 90% (Fig. 2.1). South Africa has also shown a rapid increase in the rate of recycling PET bottles (Fig. 2.2). Part of this increase in recycling rates is due to the enactment and strict implementation of recycling laws that specifically target PET bottles. In fact, Germany was the first country to implement laws on reuse and recycling of PET bottles as early as 1991 through an ordinance (BGBl. Nr. 516/1990 ST0206) [3]. In addition to countries, major beverage companies such as Coca-Cola and Evian have pledged to use at least

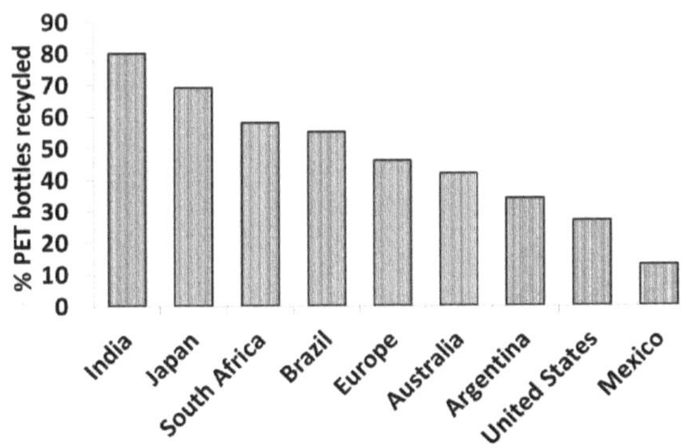

Figure 2.1 Percentage of PET bottles recycled in various countries ([1]; https://www.packaging-gateway.com/projects/mpacts-pet-recycling-plant-gauteng/).

Figure 2.2 Increase in the rate of PET bottle recycling in South Africa (http://www.globalpsc.net/pet-recycling-growth-in-south-africa/).

50% of bottles from recycled PET. Recent reports suggest that Evian has pledged to use 100% recycled PET bottles (http://www.evian.com/en_us/our_commitments_for_the_environment/taking_action_on_our_bottles_and_our_co2_emissions/life_cycle_of_an_evian_bottle/; https://www.forbes.com/sites/uhenergy/2018/03/12/plastics-recycling-could-the-future-be-in-india/#7a1046642e18).

PET bottles and other plastics are reprocessed into various products including the original form (Fig. 2.3). However, the extent and form of recycling are mostly dictated by the prevailing regulations and economics. For example, nearly 43% of PET bottles are recycled into fibers and only about 25% into food and beverage bottles in the United States compared to 34% recycling into new bottles in Germany (Figs. 2.4 and 2.5).

In addition to difficulties in collection and processing, presence of various other additives (impurities) in PET makes it difficult to reprocess. Some of the common additives used and contaminants found in PET bottles are listed in Table 2.1. Although these additivities are necessary during processing, they create problems during use and also recycling.

The European Union for Food Safety council evaluated a process of recycling of PET bottles called PET-M. It was found that this process was capable of decontaminating more than 91%−100% of

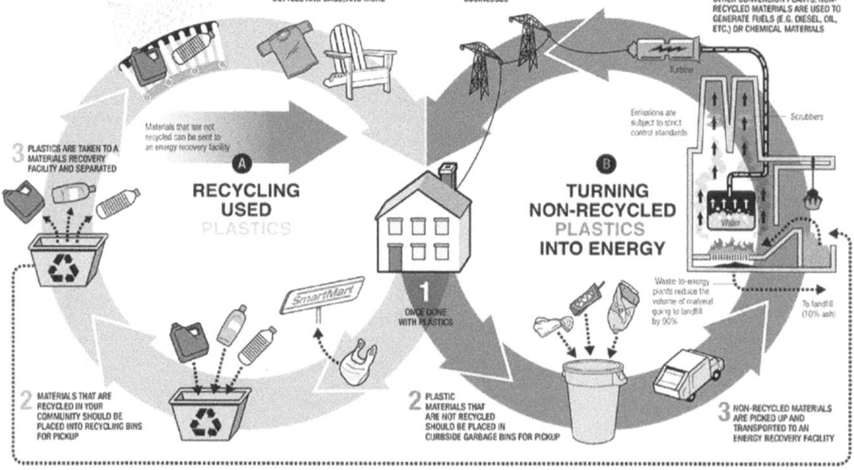

Figure 2.3 Various possibilities of reprocessing PET bottles and other plastics (https://www.plasticpackagingfacts.org/resources/diagram-showing-traditional-waste-energy-facility-recycling-energy-recovery-work-together/diverting-plastics-from-landfill-a-two-pronged-approach-pdf-2/).

Figure 2.4 Conversion of PET bottles into various forms in Germany (http://www.packagingtoday.co.uk/news/newsgermans-recycle-more-and-more-effectively-success-in-pet-bottle-recycling-according-to-study-5699219).

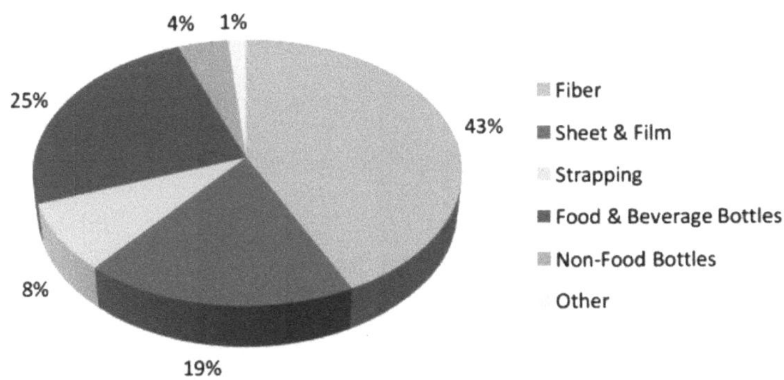

Figure 2.5 Percentage of PET bottled recycled into various products in the United States (https://www.plasticsrecycling.org/images/pdf/resources/reports/NAPCOR-APR_2016RateReport_FINAL.pdf).

contaminants. The panel recommended that the process was suitable for PET bottle to bottle recycling but suggested that the nonfood container fraction of the PET processed should be less than 5%. Tables 2.2 and 2.3 provide the decontamination efficiency for various chemicals after the PET-M process [4]. These surrogates chemical levels should not be exceeded either during prolonged storage of the component or accumulated during various levels of recycling. US FDA recommends the following surrogate chemical levels in PET bottles intended by recycling.

(https://www.fda.gov/Food/GuidanceRegulation/GuidanceDocuments RegulatoryInformation/ucm120762.htm#chem). It has also been suggested that the PET bottles tested for surrogates should be in flake

Table 2.2 Allowable Contaminants (Surrogates) Levels in PET Bottles Using Various PET Recycling Processes

Surrogates	Decontamination Efficiency (%)	C_{res} (100%) (mg/kg PET)	C_{res} (60%) (mg/kg PET)	C_{mod} (mg/kg) PET
Toluene	99.3	0.021	0.013	0.09
Chlorobenzene	99.4	0.018	0.011	0.10
Chloroform	>99.8	<0.006	<0.004	0.10
Methyl salicylate	>99.5	<0.015	<0.009	0.13
Phenylcyclohexane	96.2	0.110	0.066	0.14
Benzophenone	91.1	0.267	0.160	0.16
Methyl stearate	93.5	0.200	0.120	0.32

Table 2.3 Permissible Levels of Surrogates That Can Be Sorbed Onto PET Bottles After Exposure for 365 Days at 25°C [5]

Surrogate	Sorption (mg/kg)
Chloroform	4860
Toluene	780
Lindane	750
Tetracosane	154
Benzophenone	49
Copper (II) 2-ethylhexanoate	49
Methyl salicylate	200
Tetracosane	154
Lindane	750
Methyl stearate	150
Phenylcyclohexane	390
1-Phenyldecane	170
2,4,6-Trichloroanisole	1100

form rather than intact bottles. FDA has developed mathematical models to predict the sorption of surrogates onto PET over time. Excellent correlation has been found between experimental and actual data [6].

Most developed countries including United States, Germany, United Kingdom and Sweden have provided financial incentives to promote recycling of PET bottles. For instance, states in USA provide an incentive ranging from 5% to 10% per bottle for recycling. Similarly, Austria imposes a voluntary recycling fee of U.S. \$0.02–\$0.08 on one-way containers. Among the PET bottles, it has been suggested that about 50% of all PET bottles are used as water bottles. Extent of recycling has also been linked to the prevailing laws and their implementation. In states where the laws were implemented strictly, recycling rates of PET bottles were more than twice that in states where laws were implemented less stringently (http://www.nber.org/digest/may10/w15585.html; http://www.recyclingtoday.com/article/napcor-apr-2016-report-pet-recycling/).

According to the US regulations, PET recycling facilities should follow the following regulations [7]. Grinding systems called cyclones used to regrind the PET should follow OSHA requirements for dust exhaust and filtration systems. Similarly, transport of PET for recycling is restricted according to the Department of Transportation regulation. During transportation, the type, size, and density of the PET bales are subject to restriction. Depending on the process used to collect PET bottles, the type and specification of the bale changes. Generally, there are three types of PET bottle bales. First is the "soda bales" which contain only the PET carbonated beverage bottle, another category of bales that contain soda bottles and custom PET containers and are called as "curbside bales". Third category of bales refers to those containing only custom PET bottles and called as "custom bales". Typically, a bale is prepared with a density between 15 and 18 lbs/ft^3 and weigh between 525 and 630 lbs.

Various terminologies and methods are used to recycle and recover PET and other synthetic polymers (Table 2.4). PET bottles are generally recycled into fibers, sheets, bottles and bottles to bottles (BtoB) (http://www.petbottle-rec.gr.jp/english/design.html). The conditions of processing are different depending on the intended application. Evaluation criteria's differs for each type of product that is intended to be recycled (Tables 2.5–2.8).

Table 2.4 Terminology of the Different Recycling and Recovery Processes [2]

ASTM D 5033	ISO 15270	Others
Primary recycling	Mechanical recycling	Closed loop recycling
Secondary recycling	Mechanical recycling	Downgrading
Tertiary recycling	Chemical recycling	Feedstock recycling
Quaternary recycling	Energy recycling	Valorization

Table 2.5 Evaluation Criteria for Processing PET into Fibers

Evaluation Items		Evaluation Details	Unit	Specifications
Processability	Filtration	Increase in filtration pressure after 3 h of continuous running	kg/cm^2	20 >/=
	Spinning	Yarn breakage at spinning during 3 h of continuous run	Times	1
	Drawing	MDR	%	90 >/=
		Yarn breakage at drawing	Times	1
	Yarn Quality	Toughness		20 >/=
		Shrinkage in boiling water	%	80–120
		Dry shrinkage rate	%	80–120
		Color (Δb)		2 >/=
		Thickness	%	80–120
		Tensile strength	%	80–120
		Tensile elongation	%	30 ± 5
		Intrinsic viscosity		0.55–0.7
	Dyeability	Dye uptake rate	%	90 >/=
		Spots		No spots
		Colorfastness to washing	Class	4 >/=

Several countries have agencies that form rules and regulate the use, disposal and recycling of PET bottles. Some of the major agencies or organizations involved in PET recycling are listed in Table 2.9 below lists the various agencies that are involved in setting and monitoring regulation on recycling PET bottles. The EFSA stipulates that PET bottle recycling should contain less than 5% of nonfood plastics.

The European PET Bottle Platform (EPBP) has developed a few tests where individuals can evaluate the recyclability of PET bottles. Oven test which tests the presence of impurities when heated to 220°C, sink float separation test where components heavier than PET are separated,

Table 2.6 Evaluation Criteria for Sheets

Evaluation Items		Evaluation Details	Unit	Specifications
Processability	Filtration	Increase in filtration pressure after 2 h of continuous running	kg/cm^2	10 >/=
	Sheet molding	Sink or flow marks		No marks
		Number of fisheyes	Per m^2	2 >/=
	Thermoforming	Contour		Similar to reference
		Whitening		No whitening
	Sheet properties	Haze	%	3 >/=
		Color (Δb)	%	2 >/=
		Impact strength	%	85 >/=
		Tensile yield strength		85 >/=
		Fracture point	%	85 >/=
		Tensile modulus	%	85 >/=
		Spots		No spots
		Colorfastness to washing	Class	4 >/=

glue separation and label adhesive test are recommended to understand the recyclability and potential problems that may arise when PET bottles are recycled (http://www.epbp.org/page/8/downloads#downloads_39). Plastics Forming Enterprises in Amherst, NH, USA and Europe Parc Technologique in Switzerland are two EBPB accredited laboratories that certify the recyclability of PET bottles. The PET bottles and materials tested by them are also classified as fully compatible, limited compatible and noncompatible for easy identification during recycling.

Table 2.7 **Evaluation Criteria for Bottles**

Evaluation Items		Evaluation Details		Unit	Specifications
Processability	Injection molding	Appearance of the preforms			Similar to reference material
	Blow molding	Appearance of the bottles			Similar to reference material
	Thermoforming	Contour			Similar to reference
		Whitening			No whitening
	Bottle properties	IV retention		%	90 > / =
		Acetaldehyde		%	110 > / =
		Bottle dimensions	Brimful capacity	%	1.0 > / =
			Height	%	1.50 > / =
		Thickness distribution	Average wall thickness	%	90 > / =
			Thickness evaluation	%	110 > / =
		Color (Δb)			3 > / =
		Haze		%	5 > / =
		Tensile yield strength		%	90 > / =
		Fracture point		%	85 > / =
		Drop impact strength		%	85 > / =
		Spots		Times	1 > / =
		Whitening from drop impact			No whitening

Table 2.8 Evaluation Criteria for Bottle to Bottle PET Recycling

Evaluation Items	Evaluation Details	Unit	Specifications
BtoB processed pellets	IV increase ratio	%	Rank A: 3 >/=, 120 >/+ Rank C: 90>, 120 <
Injection molding plates	Flow marks		Rank A: No flow mark Rank C: Flow mark
	Sink		Rank A: No sink Rank C: Sink
	Haze	%	Rank A: 5 >/= Rank C: 5 <
	b^*		Rank A: 12 >/= Rank C: 12 <
	Δb^*		Rank A: 5 >/= Rank B: 2 < Δb, /= 5 Rank C: 5 <

Table 2.9 Agencies That a Responsible for Implementing Recycling and Reuse Legislations in Various Countries

Country	Agency
Japan	The council for PET bottle recycling (http://www.petbottle-rec.gr.jp/english/index.html)
European Union	European Food Safety Authority (EFSA) Panel on Food Contact Materials, Enzymes, Flavorings and Processing Aids, Article 4 of Regulation (EC) No. 282/2008 – (EC) No. 282/2008-Guidelines on Submission of a Dossier for Safety Evaluation by the EFSA of a Recycling Process to Produce Recycled Plastics Intended to be Used for Manufacture of Materials and Articles in Contact with Food – The commission of the European communities-COMMISSION REGULATION (EC) No. 282/2008 (http://data.europa.eu/eli/reg/2008/282/oj) – Directive 94/62/EC EU regulation on plastic and recycling
India	Bureau of Indian Standards: IS 14535: 1998: Indian Standard for Recycled plastics for the manufacturing of products—Designation IS 14534: 1998: Guidelines for the recovery and recycling of plastic waste

Table 2.9 Agencies That a Responsible for Implementing Recycling and Reuse Legislations in Various Countries—cont'd

Country	Agency
Australia	Australia Packaging Covenant, http://www.environment.gov.au/protection/national-waste-policy/packaging-covenant
France	Proposition 2015-922 restricts the sale of disposable plastic goods including PET bottles from 2020
Mexico	Management Plan for PET Packaging Waste (Plan de Manejo de los)
	Residuos de Envases de PET (September 2002)
China	GB 16487 (2017): standard for imported scrap materials
United Kingdom	Waste Regulations 2011
United States	SB-458 Beverage container recycling (Legislation in California)
	FDA CFR 170.39
	Points to Consider for the Use of Recycled Plastics in Food Packaging: Chemistry Considerations' (FDA 1992)

2.2 Conclusions

It is surprising and rather unfortunate that there are a very limited number of legislations on PET bottle recycling. Despite PET bottles being one of the most widely used commodity products and incentives being provided for recycling the bottles, the extent of recycling is limited except in a few countries. Laws on recycling only specify the levels of contaminants to be removed or present and in some instances, the amount or recycled PET allowable is also restricted. Further regulations on mandatory recycling, properties of the recycled bottles and processing information are necessary to enhance the level of recycling.

Acknowledgments

Authors express their thanks to the Center for Incubation, Innovation, Research and Consultancy for their financial support. Narendra Reddy is thankful to the Department of Biotechnology, Ministry of Science and Technology, Government of India for funding through the Ramalingasami Re-entry Fellowship.

References

[1] T.M. Coelho, R. Castro, J.A. Gobbo, PET containers in Brazil: opportunities and challenges of a logistics model for post-consumer waste recycling, Resour. Conserv. Recycl. 55 (3) (2011) 291–299.
[2] J. Hopewell, R. Dvorak, E. Kosior, Plastics recycling: challenges and opportunities, Philos. Trans. R Soc. London B: Biol. Sci. 364 (1526) (2009) 2115–2126.
[3] E. Schwanse, Recycling policies and programmes for PET drink bottles in Mexico, Waste Manage. Res. 29 (9) (2011) 973–981.
[4] E.F.S. European Food Safety AuthorityAuthority, Scientific opinion on the safety evaluation of the following processes based on EREMA Basic technology used to recycle post-consumer PET into food contact materials' Coveris' and'Envaplaster', EFSA Journal J 13 (4) (2015) 4065.
[5] T.H. Begley, T.P. McNeal, J.E. Biles, K.E. Paquette, Evaluating the potential for recycling all PET bottles into new food packaging, Food Addit. Contam. 19 (S1) (2002) 135–143.
[6] A. Baner, J. Brandsch, R. Franz, O. Piringer, The application of a predictive migration model for evaluating the compliance of plastic materials with European food regulations, Food Addit. Contam. 13 (5) (1996) 587–601.
[7] D.J. Hurd, Best practices and industry standards in PET plastic recycling, Washington State Department of Community, Trade and Economic Development's Clean Washington Center, Seattle, WA, 98121, Contract # 597-220-028.

Further Reading

http://www.petcore-europe.org/news/pet-collection-and-recycling-rates-europe-significantly-increased-2016.
https://napcor.com/reports-resources/environmental-impact/.
https://www.fda.gov/Food/GuidanceRegulation/GuidanceDocumentsRegulatoryInformation/ucm120762.htm#chem.
R. Dvorak, E. Kosior, J. Fletcher, Improving food grade rPET quality for use in UK packaging, Project code: IMT003-001, Research date: November 2011–March 2012, July 2013.
H. Zhang, Z.-G. Wen, The consumption and recycling collection system of PET bottles: a case study of Beijing, China, Waste Manag. 34 (6) (2014) 987–998.

3 Materials Recovery, Direct Reuse and Incineration of PET Bottles

Cintil Jose Chirayil[1], Raghvendra Kumar Mishra[2] and Sabu Thomas[2]
[1]Department of Chemistry, Newman College, Thodupuzha, Kerala, India, [2]International and Inter University Centre for Nanoscience and Nanotechnology, Mahatma Gandhi University, Kottayam, Kerala, India

3.1 Introduction

Plastics are inexpensive, light in weight as well as durable materials that may conveniently be molded into a variety of products which search for employed in several applications. For this reason, the development of plastics materials has increased substantially over the last few years. Up-to-date quantities of their consumption and disposal generated a range of environmental problems. At least 4% of global coal as well as oil generation, a nonrenewable material, which is employed in form of feedstock for plastics materials and an extra 3%−4%, is used to provide us energy with their manufacturing [1]. A critical section of plastic materials formulated each year is employed which could apply in the role of parts and components of packaging or various other minor things that are thrown away within every year of manufacturing. The two scientific studies alone recommend use; current application of plastics is undoubtedly not environmentally friendly. Additionally, on account of the durability of the polymers involved, considerable levels of scrapped end-of-life plastics are accumulating in form of waste in landfills as well as in natural habitats globally. Reusing is the main guidelines, right now which is readily available for reducing these types of influences and gives possibly one of the most dynamic areas when we look at the plastics industry nowadays. For instance, reusing offers opportunities to minimize oil usage, skin tightening and emissions in addition to degrees of waste demanding disposal [2−4]. The intention of this chapter is actually to evaluate the recovery, direct reuse as well as incineration of polyethylene terephthalate (PET) bottles recycling system, wherein more widely used bottles are reprocessed into mutually fiber as well as bottles, additionally to evaluate the recycling with a virgin. On account of, PET is now the most reliable

Recycling of Polyethylene Terephthalate Bottles.
DOI: https://doi.org/10.1016/B978-0-12-811361-5.00003-1

product packaging material globally for water together with soft drinks bottles. The reason for this progression is the excellent material features of the PET material, especially its mechanical properties along with the very low weight of the bottles in comparison with glass bottles of the similar filling volume. As compared to supplementary packaging polymers, PET has additionally an excellent clarity as well as beneficial barrier properties towards moisture and oxygen. As a result, PET has predominantly replaced glass in the role of packaging material, nevertheless also metal cans. In these days, PET bottles can be used for soft drinks, drinking water, energy drinks, ice teas as well as for more delicate drinks such as beer, wine and also juices [5]. The variety of PET bottles for beverages is still expanding all over the world. For decades the reusing postconsumer product packaging materials into totally new packaging application were attainable just for glass bottles and consequently jars just like for metal cans. For product packaging polymers, a reusing of postconsumer packaging materials into direct food contact packaging applications was challenging, as a result of the inadequate understanding about contamination of the packaging polymers throughout very first use or recollection. Furthermore, the decontamination efficiencies of recycling techniques were generally unidentified in so many cases. Hence, the evaluation of the food law compliance level of the postconsumer recyclates was unachievable for decades. The major problem was that ingredients from the drinks, as well as harmful compounds from possibly exploited containers for storage of domestic cleaners or garden chemicals, is often absorbed into the polymer. In the event that these types of absorbed postconsumer toxins are not taken away from the packaging polymer in the course of recycling, they may migrate into food items from the recyclate having packaging materials [6]. Additionally, chemicals that are not accepted for foodstuff packaging materials can go into the foodstuff packaging stream when nonfood packaging materials of the identical polymer form were as well recollected and reused along with the food items packaging materials. Due to the exceptionally inert aspect of PET, recycling technologies have been established to develop a bottle-to-bottle recycling of postconsumer PET bottles [7].

The end result of altering of reprocessed PET pellets which will make bottles or fiber. Bottle-to-fiber recycling offers environmental impact reduction. The understanding explanations show that the reprocessed PET content in a reprocessed bottle is certainly not important into the as a whole environmental performance [8–11]. Waste plastics items are restored when they are redirected from garbage or littering. Plastic product packaging is particularly identifiable in form of the bottle because of the lightweight feature of both flexible as well as rigid plastics.

Developing the items by reusing, recovering or even remanufacturing can certainly help in lesser items transporting into the waste goods stream. When material circulates into the waste stream, reusing and recycling is the approach to reestablishing the material to generate a distinct product. Regarding materials including plastics, the idea of recovery may be enhanced to contribute energy recovery, wherein reality the calorific value of the materials is employed by controlled combustion in form of a fuel, although this leads to an inferior entire environmental performance as compared to material recovery since it will not reduce the necessity of new (virgin) material [2,12–14]. This thought can be on the four basic approaches in waste control. In the order of reducing environmental desirability: reduce, reuse, and recycle the materials and recover the energy, while landfill offered least desired control approach [15]. Additionally, it is truly easy for the exact same polymer to stream by multistages, for example, fabrication into a reusable container, which once going into the waste stream is gathered and recycled into useful products, whenever becoming waste with its turn, is recovered for energy. PET bottles have seen sudden progress since the 1970s, the means of blow molding was introduced [16]. Nowadays, bottle grade PET is the most relevant packaging plastics, and it is probably the most widely used plastics. Almost all of the recycled PET (r-PET) flakes were changed into fibers. PET bottles are ubiquitous in our day-to-day lives. r-PET can be utilized in a wide range of applications [10,17,18]. These plastic bottles are obtained by the garbage collector or waste dealers, who use men and women to separate, sort and additionally trade it to large companies or recyclers. The hats, opening rings, product labels (non-PET components) are extracted, and the plastic bottles are chopped, cleaned, and supplied as “cleaned flakes.”. These cleaned flakes are then utilized to produce polyester fiber, which is utilized as filling material for cushions, pillows, and transformed to fabrics for applying in garments, furniture, etc., as shown in Fig. 3.1 [11,19].

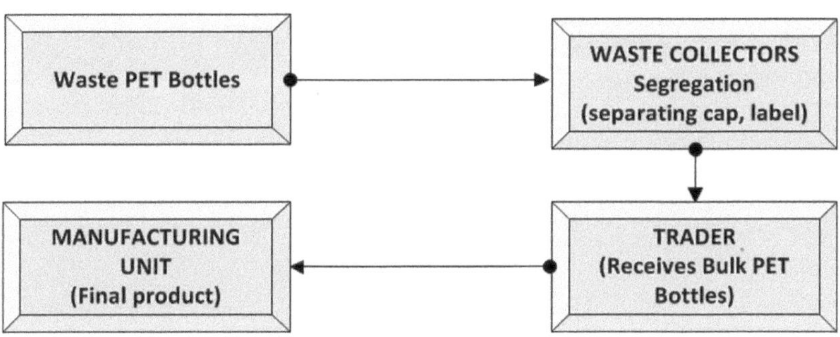

Figure 3.1 Recycling of waste PET bottle or plastics.

3.2 An Overview of Polyethylene Terephthalate Bottle Management

In general language, waste plastics are recovered when they are distracted from landfills or even littering [2,20]. Developing items to permit reusing, restoring or even remanufacturing have an impact on the goods going the waste stream. Recycling is the procedure for utilizing recovered material to build a new type of products [21].

There is considerable opportunity for reuse of plastics in the transport products, as well as feasible reuses or remanufactures consumer products such as vehicles and electronic equipment. It is recognized on an industrial scale with reuse of containers as well as pallets in haulage. The certain transition of single-use plastic carrier bags to reusable bags has additionally been observed, both as a result of voluntary behavior change programmes, such as Australia and also as a result of legislation, including the plastic bag levy in Ireland, or perhaps the current banning of lightweight carrier bags in some Asian countries [14,22,23]. A brief PET life cycle is shown in Fig. 3.2.

3.2.1 Recycling of Polyethylene Terephthalate Bottle

Plastics recycling need to be taken into account in almost any plastic waste-management plan. Along with decreasing the volume of waste disposal in landfills, it may also considerably lead to the preservation of

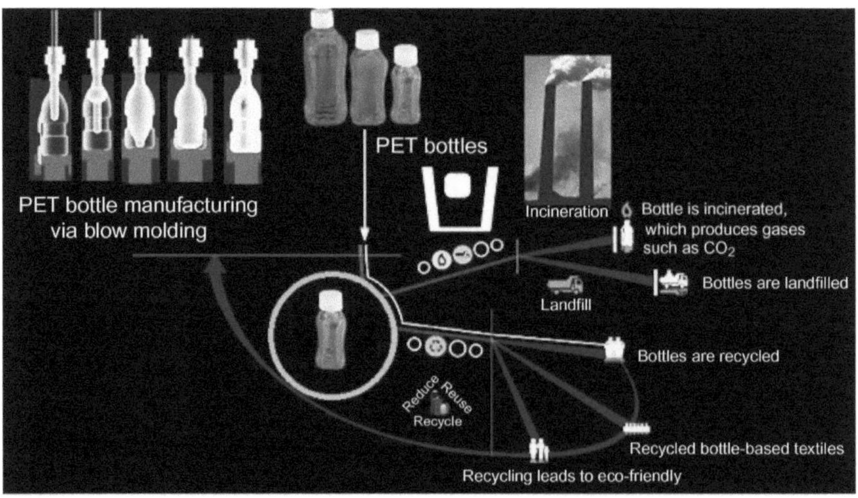

Figure 3.2 PET bottles life cycles.

raw petrochemical items, along with energy savings. Authors have discussed that there are several technological as well as industrial limitations that presently restrict the complete as well as effective recycling of plastic waste items into valuable products, and these are [20]: (1) polluting of plastic waste items with other elements, for example, dust together with metals which can deteriorate the tools employed in the reprocessing of the waste; (2) plastic blends are generally insoluble and develop discrete phases within a continuous phase; (3) plastics are not uniform materials like aluminum or paper, however, are comprised of a great number of qualities with various molecular structures and features, and each and every plastic ingredient in an integrated waste possesses a distinct melting tendencies, rheology, and thermal stability; (4) plastic waste feedstock is not generally homogeneous with time, and (5) plastic waste items have a comparatively low density [24]. As a result, they are generally compacted or ground-up before transportation to decrease delivery costs. We can find a lot of people mixed up in PET recycling business, beginning with the waste collectors to small and medium dealers, recyclers, and producers who make use of r-PET for assorted products. There are many different ways that PET bottles arrive at the waste stream for recycling. Municipal solid waste is the one primary source. Waste collectors also go right to the sources (railroad, stations, national/international airports, cafes, resorts) to get. PET bottles are separated from the dry waste and discarded, there clearly was a top likelihood that they can be picked up and recycled. There are lots of factors that set PET bottles aside from other plastic waste. PET bottles can be differentiable, therefore very easy to choose out, format, store, etc. As well as weight per bottle (empty) is fairly higher for specific waste collectors to spend resources to get them (in comparison with quite minimal weight, smaller size waste, prefer chocolate wrappers). We can find production units which make use of r-PET (washed flakes) as raw material [11,25,26]. In sorting procedures, people sort the waste material into polyethylene (PE), polypropylene (PP), milky PP, high-density polyethylene (HDPE), low-density polyethylene (LDPE), PET, etc. This is accomplished commonly by visible assessment, by exposing the plastic stuff to a few improvised checks like the look, twisting, snapping, using solvents, etc. The various plastics are separated and sold to various traders. Every single PET bottle is inspected strictly by using manual assessment. Polyvinyl chloride (PVC) labels pose a certain issue, while they could severely risk the fiber manufacturing process. PVC and PET have an equivalent density, thus it is hard to separate your lives through flotation, but PVC and PET have own interior viscosities that make it extremely tough to process PET polluted with PVC. All PVC labels (identified by sight) are eliminated [27]. Bottles made of other plastics are eliminated. When all this work is completed, the bottles are broken which will make lightweight and small size bundles. Baling assists us to ship PET bottles efficiently by rendering it feasible ship in a full-scale truck [11,25,26,28]. Numerous PET bottles, if they get to the dealer, may include nontrivial quantities of residual liquids (water, juices, other drinks, liquids kept in the bottle). Some of those liquids are removed by using the sorting process. A number of the liquids are crushed out by using the baling process, all of that causes a loss in weight regarding the PET bottles-shortage could be the term used to denote this loss. Trucks transporting bales of PET bottles get to the recyclers. The bales are opened additionally the sorting, cleaning process starts yet again. The bottles are positioned by using different steps of manual, semiautomatic and automatic sorting, cleaning procedures. All PVC labels, bottles are eliminated. Any kind of metal element which may possess (such as caps, rings, springs found in spray bottles, etc.), those are faithfully eliminated. The bottles are shredded, in addition to plastic (non-PVC plastics such as PE, PP) elements are removed by flotation [11,29,30]. The flakes are afflicted by further washing, drying, consequently, they are available (cleaned flakes) to be shipped to the production units. The cleaned flakes tend to melt and spun into fibers of varied qualities, and which will make PET sheets, monofilament, strap, etc. Recycling process would operate in the organized sector. You can find a lot of units, plants that generate cleaned flakes of varied features and qualities which can be yet again fed into other units that generate a selection of polyester/PET products. Making use of r-PET may be spotted because of the slight yellow tinge, this kind of PET bottles recycling is shown in Fig. 3.3 [11,29,30].

3.2.1.1 Collection

The number of plastic wastes is accomplished by delivering-schemes or by using waste collector. Deliver-plans tend to low collection rates whenever we look at the insufficient highly dedicated public or deposit-refund plans which employ an essential economic inspiration to participate. Because of this, the typical trend is in fact for the number of recyclable materials by using waste collector. The majority waste collections are the paper, glass, board, metals as well as plastic

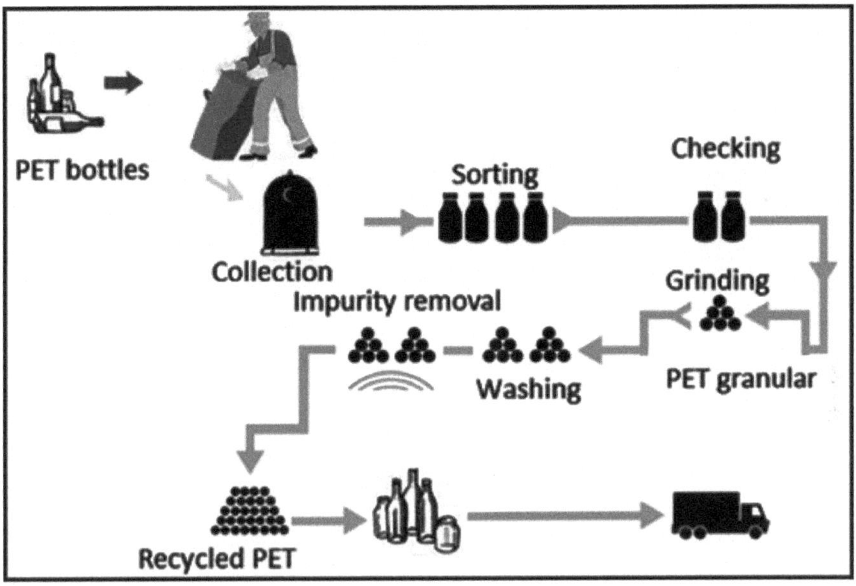

Figure 3.3 Recycling of PET bottles.

containers. Nevertheless waste collection strategies have now been pretty reliable at recovering plastic bottle packaging from residences, an entirely consumption generally only 30%–40% of postconsumer plastic material bottles are recovered [31–33].

3.2.1.2 *Sorting*

Sorting of rigid recyclables is performed via either automated or manual methods. Automatic presorting is generally adequate due to a plastics stream sort-out from glass, metals as well as paper [13,34]. Generally, very clear PET and unpigmented HDPE dairy bottles are favorably identified as well as segregated out from the stream. These kinds of systems generally take advantage of Fourier-transform near-infrared spectroscopy for polymer structure study and additionally employ optical color identification camera devices to sort-out the streams [13,34]. Optical sorters help to distinguish in between clear, dark blue, light blue, green as well as various colors of the PET packing containers. Sorting efficiency could be enhanced by utilizing sorting in series and multiple detectors. However, the developments of new and emerging technologies, for example, ballistic separators, sophisticated hydro cyclones and air-classifiers may raise the capacity to recover postconsumer flexible packaging [35–37].

3.2.1.3 Cleaning

Plastics are frequently changed into flakes and consequently washed to eliminate food items residues, pulp fibers as well as adhesives. The recent development of cleaning plants makes use of only 2–3 m^3 of water per tonne of material, approximately one-half of that of earlier equipment. Innovative alternative technologies for the reduction of organics as well as surface contaminants from flakes consist of "dry-cleaning," which cleans surfaces via friction without the use of water [38–40].

3.2.1.4 Final Separation

Just after size reducing, separation strategies may be employed. Sink or float separation in water can efficiently to separate polyolefins from PVC, PET and polystyrene. Usage of various media makes it possible for separation of polystyrene from PET, but PVC is not taken out of PET this way because their density varies concurrence. Some other separation strategies such as air elutriation could also be used for the removal of low-density films from denser soil plastics, e.g., in the removal of brands labels from PET flakes [41–43]. Technologies for the decreasing PVC pollutants in PET flake consist of Raman emission spectroscopic detectors to allow flake ejection and implementing a variety of electrostatic properties. For the PET flake, thermal kilns could be used to precisely break down minor quantities of PVC pollutants, as PVC transforms black on heating, enabling color-sorting [44–46]. Different techniques are present for the flake-sorting, however, conventional PET-sorting systems are mainly limited to separating: (1) colored flakes from clear PET flakes and (2) materials with different physical properties, for example, density from PET. Emerging techniques, for example, laser-sorting systems can help to eliminate other pollutants such as silicones and nylon [38,47–49]. Laser-sorting' makes use of emission spectroscopy to distinguish polymer varieties. These systems will likely considerably enhance the capacity to separate complex mixtures as they possibly can execute as much as 860,000 spectra/s and can examine every person flake. They usually have the bonus they can be employed to sort completely different plastics which can be black, a problem with conventional automatic systems. The effective use of laser-sorting systems probably will enhance separation of automotive plastics. These systems well the power to separate polymer by form or grade and certainly will well separate polyolefin materials, for example, PP from PE [50–52].

3.2.2 Types of Polyethylene Terephthalate Recycling

Because of the excellent properties of the PET including corrosion resistance, low density, superior strength, and also easy to use and design, PET application is becoming considerably more than the application of aluminum or other metals. The PET is processed by a number of techniques resulting in a variety of plastic items. They are applied for a certain use various compounding, operational condition, artificial additives, fillers, as well as reinforcements. Various processing methods are employed at this point to create plastic products with the cheapest price range. The recycling, as well as incineration, is the common facets of recovery methods in the matter of PET [53]. The incineration offers certain complications such as the generation of toxic gases along with the residue ash which has the lead and also cadmium. The recycling provides positive aspects, for example, lessening of environmental problems and consequently preserving both materials as well as energy. The plastic can be degraded in the natural environment by four mechanisms: photodegradation, hydrolytic degradation, thermo-oxidative degradation, and biodegradation by microorganisms. The all natural degradation of plastic starts with photodegradation because of the ultraviolet light from the sun that supplies the activation energy needed to start the involvement of oxygen atoms in the polymer, resulting in thermo-oxidative degradation. Within this step, the plastic turns into brittle and its fracturing into small pieces until the polymer chains attain fully low molecular weight to be metabolized by microorganisms, microbes transfer the carbon of the polymer chains to carbon dioxide or even the integrate it into biomolecules [54–56], however, this method will require a minimum of five decades. Therefore, an option for these types of issues will be the recycling, since the majority of commodity plastics are reasonably stable, generating monomer recovery inadequate. The term for plastics recycling is complex together with tricky because of the variety of recycling and also recovery strategies. These usually are comprised of four kinds: primary is a mechanical reprocessing into a product with similar properties, secondary is a mechanical which is used for reprocessing into merchandise having lesser properties, tertiary is used to recovery of the chemical component and quaternary is applied for recovery of energy. Primary recycling is in fact referred to as closed-loop recycling and second recycling as downgrading. Tertiary recycling is possibly known as chemical or feedstock recycling and is applicable as soon as the polymer is depolymerized to its chemical composition. Quaternary recycling is energy recovery, energy from waste [57–59].

3.2.2.1 Primary Recycling

The most common method is defined by the primary recycling as a result of their ease of use and affordable. This method pertains to the reuse of goods in their initial structure. The downside of this process is characterized by the restriction on the number of cycles for each [60].

3.2.2.2 Secondary Recycling

This method is likewise known as mechanical recycling. Within this procedure, only the thermoplastic polymers can be employed, as they can be remelted as well as reprocessed into finish goods. The mechanical recycling is not going to entail the modification of the polymer throughout the process. This technique is symbolized by a physical process, wherein the plastic waste items are shaped by chopping, shredding or cleaning into granulates, flakes or pellets of suitable quality for production, after which melted to create the brand new item by extrusion. Additionally, the reprocessed material is usually melt mixed with virgin material to achieve superior results. After the plastic is classified, washed, dehydrated and after that precisely processed into finish goods, the volume of the waste plastic material will be significantly diminished. The drawbacks of this technique relate to the heterogeneity of the solid waste materials along with the lack of product's features in each cycle which happens because of the low molecular weight of the recycled resin. It is mainly because of chain-scission reactions due to the existence of water as well as acidic contaminants, in order to prevent reducing molecular weight extensive drying is advisable [61–63].

3.2.2.3 Feedstock or Tertiary Recycling

This method is also termed as chemical recycling. This method is used for mechanical recycling in the role of a complementation. Chemical recycling is known as the process wherein polymers are chemically transformed to monomers or to a certain extent depolymerized to oligomers via a chemical reaction. The developed monomers are employed for fresh polymerizations to recreate the primary or a relevant polymeric product. This process can convert the plastic material into smaller molecules, appropriate for employ as feedstock material beginning with monomers, oligomers, or mixtures of other hydrocarbon materials. Hydrogenation, glycolysis, methanolysis, gasification, pyrolysis, hydrolysis, chemical depolymerization, photodegradation, ultrasound degradation, degradation

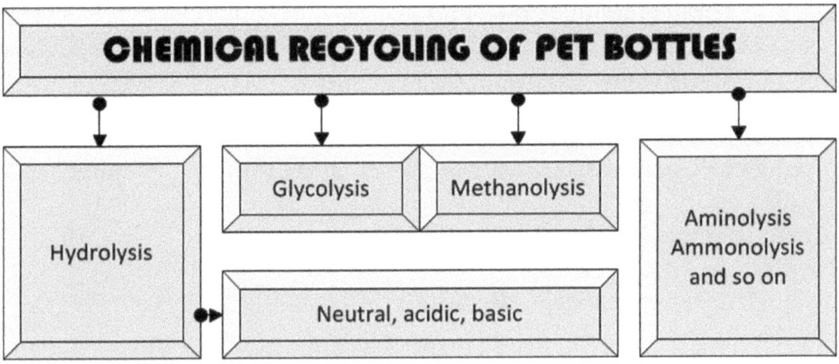

Figure 3.4 Chemical recycling of PET.

in the microwave reactor and so forth can be employed for decomposition of polymers into monomers [53,64,65]. The chemical recycling is not completely established for this purpose, just a few manufacturers are actually working on it due to the fact this process requires more capital as well as the specialist. At this moment, several techniques are under examination; as an illustration, the gasification, as well as pyrolysis process is under considerable investigation to develop the appropriate conditions, and the methods which have achieved industrial growth at this point are glycolysis and methanolysis. Polyethene terephthalates are cleaved by certain reagents, for example, water in the case of hydrolysis, acids in the case of acidolysis, glycols in the case of glycolysis or alternatively alcohols for alcoholysis [66–68]. Based on the reagent applied, various goods are achieved. Among the almost all recycling methods, the only one satisfactory as per the concepts of environmentally friendly development is chemical recycling, because it contributes to the production of the monomers from which the polymer is manufactured. The recycled PET is mostly employed as fibers, films, foams, sheets, bottles, etc. Various types of chemical recycling of PET are discussed in Fig. 3.4.

3.2.2.4 Quaternary Recycling

This process represents the recovery of the plastic's energy content. The most reliable approach to bring down the volume of organic materials, the recovery of energy is achieved by incineration. This process is an efficient option since it produces significant energy from polymers; however, it is not ecologically suitable due to the health hazard from airborne poisonous materials. In this manner, various authors have performed the study on the reusing or recycling of PET as well as PET bottles, as mentioned in Table 3.1.

Table 3.1 Application of Waste PET

Type of Polyethylene Terephthalate Studies	Type of Use	Reference
Polyethylene terephthalate bottles	Reuse as well as recycling	[69]
Polyethylene terephthalate bottles	Recycling	[70]
Polyethylene terephthalate bottles	Recycling	[71]
Polyethylene terephthalate bottles	Recycling	[72]
Polyethylene terephthalate bottles	Recycling	[73]
Polyethylene terephthalate bottles	Recycling	[17]
Polyethylene terephthalate bottles	PET bottle-to-fiber recycling	[71]
Polyethylene terephthalate bottles	Dimethyl terephthalate production from recycled PET	[74]

3.2.3 Materials Recovery From Polyethylene Terephthalate Bottles

The polyester has to be recycled is converted into an oligomer by introducing ethylene glycol or supplementary glycols throughout thermal treatment. The objective, as well as the benefit of in this way of handling, is the chance of filtering the mechanical deposits straightly and effectual via a gradual as well as stepwise filtration. The filtration quality of the final filtration stage possesses a well-defined influence on the level of quality of the final product [75]. For example, by the hydrolysis method, hydrolysis of the PET to purified terephthalic acid and ethylene glycol, is achieved under high pressures as well as supercritical situations. In this instance, PET-waste is straightly hydrolyzed using supercritical water steam. The filtration of crude terephthalic acid is performed by recrystallization in acetic acid/water mixtures much like purified terephthalic acid filtering [68]. Methanolysis is the recycling method that has been practiced as well as analyzed on a significant scale for several years. In this scenario, polyester waste is converted with methanol into dimethyltryptamine, under pressure as well as in existence of catalysts. After this, an effective purification of the methanolysis product is used. Lastly, the crude dimethyltryptamine has purified by using vacuum distillation [68].

3.2.3.1 Direct Reuse of Polyethylene Terephthalate Bottles

Recycling procedures with polyester are practically distinct due to the fact the manufacturing methods based on primary pellets or melt. Based on the purity of the recycled materials polyester can be employed in the majority of the polyester manufacturing techniques as a melt mixed with a virgin polymer or progressively 100% recycled polymer. This method comprises in converting bottle waste materials into flakes, by drying as well as crystallizing the flakes, by plasticizing and consequently purification, along with pelletizing [76]. A unique characteristic is: acetaldehyde and also oligomers are found in the pellets at smaller level; the viscosity is decreased in some way, the pellets are amorphous that have to be crystallized and consequently dried before additional processing. In an effort to reduce the costs, one is focusing on the direct use of the PET flakes, from the treatment of second-hand bottles, with the intention of processing an increasing amount of polyester intermediates. For the correction of the required viscosity, apart from an effective drying of the flakes, it is really essential to reconstitute the viscosity via polycondensation in the melt phase or even solid-state polycondensation of the flakes [77]. The recent PET flake transformation operations are employing twin screw extruders, multiscrew extruders or multirotation systems as well as coincidental vacuum degassing to take away moisture together with preventing flake predrying. These types of operations help the transformation of undried PET flakes without considerable viscosity reduce due to hydrolysis. Taking a look at the use of PET bottle flakes the primary percentage of approximately 70% is transformed to fibers as well as filaments. Beginning with bottle flakes, the viscosity is established by means of the degree of drying. The supplementary application of TiO_2 is essential for entire dull or semifibers yarn [78]. Rooftop covers or in road structure are made by spinning bottle flakes. With regard to high tenacity product packaging stripes as well as monofilaments, the preliminary raw material is a mostly reused material of higher intrinsic viscosity. High tenacity packaging stripes, in addition to monofilament, are produced in the melt spinning operation. Traditional recycled PET flakes are reused in the foodstuff packaging field when the recycled postconsumer PET is applied with a coating of virgin PET [79]. The virgin coating plays the role of an effective barrier and also decreases the migration of possible postconsumer pollutants from the core layer into the food items. The use of functional barriers is on the first approach to use recyclates into foodstuff packaging materials since the postconsumer recyclates do not have a direct contact with the food items. As a result, multilayer bottles or even sheets are one of the primary recyclate having product packaging architectural structures, which have been launched into the marketplace. Nevertheless, a basic challenge of functional barriers is that the foodstuff contact virgin layer may be polluted with postconsumer impurities from the core layer in the course of sheet or preform/bottle manufacturing. Because of the above the melting temperature of PET, the diffusion coefficients of impurities are extremely high, in order that the postconsumer impurities are able to migrate from the recyclate having core layer into the virgin food contact layer. As a result, the quantity of postconsumer impurities in the virgin layer of multilayer PET bottles affects the thickness of the functional barrier [77,78].

3.2.4 Incineration of Polyethylene Terephthalate Waste

The incineration of waste as a fuel releases heat energy in cement factories and also energy power plants that are used in technical devices. The resultant heat is employed for functioning systems, heating system as well as producing power. The downside of combustion of plastics is the air pollution due to the dangerous fumes discharged into the atmospheres. The incineration of waste in form of a fuel is achievable, consequently, according to licensing only as outlined by lawful specifications and technological innovation incineration plants with flue gas purification equipment together with the consistent tracking of the combustion gases released. Plastic-type waste is incinerated in certified plastic waste materials incineration plants, all other kinds of burning plastic waste are prohibited [80,81]. Typically plastic waste materials are produced by usual households. The arrival of superior selective waste assortment systems has enabled the separation of various materials together with kinds of waste. A significant duty is always to highlight the advantages of the separation of plastics, so that they turn out to be reusable and also much less harming to our environment. In the course of incineration, plastics induce long-term harm to the combustion heater in the flue systems and the resultant combustion items create a dangerous risk to both people along with the environment. The burning up of plastics is a complicated chemical operation. Based on their configuration, plastics are generally micromolecular or even macromolecular materials. Throughout plastic combustion, various phases occur, for example, warming, degradation, flashover, combustion almost all are found simultaneously [82]. Low molecular compounds are vaporized straightly in the air, and based on their wide variety, can make a combustible mixture, or oxidize in a solid manner. Macromolecular plastics need to degrade into tiny molecule compounds to start the combustion operation. Burning is combined with the production of chark, coking extent is dependent upon the situations of combustion. Two zones are usually established in the course of the combustion of nearly all plastic. The first zone is the gas evolution (pyrolysis aspect), the second zone is the chark zone (between the surface and the pyrolysis zone). The chark zone contains porous solid residues. Gases created in the course of the decomposition of the plastic composite items are incredibly hazardous. The combustion aspects of PET plastics are 500°C (ignition temperature), 22.7 MJ/kg (heat of combustion) [83]. The combustion tools as well as chimney system are intended to burn specified fuels. The change from the recommended is actually a risk as they will not produce the flue gas exhaust, and also drafts of ventilation required. The burning of household waste entails a considerable amount of flue gas production, a part of is settled in the chimney wall structure and due to the lack of built-in filtration tools, then the rest come into the natural environment. The soot accumulation in the chimney happens throughout intense plastic waste, burning up rubber derivatives. The incineration of blended waste leads to considerably of smoke. The smoke flowing by the chimney cools down constantly. As a result of the reduction in temperature, water vapor condenses on the chimney wall structure from the smoke [84,85]. The plastic substance for smoke derivatives present in the precipitating liquid is acidic. This acidic fluid destroys the chimneys internal plaster. Because of the consistently adhering flue combustion products within the narrow chimney, consequently, the temperature of the exhaust toxic gases raises. Because of this temperature rise, the interior temperature of the chimney additionally increases. After achieving the ignition temperature of the flue combustion items settled in the chimney they ignite along with a chimney fire takes place. The interior temperature is able to achieve 1000°C when burning. Taking into consideration the negative effects of toxic gases caused by the combustion of plastics, it needs to additionally be stated that the soil as well as the aquatic ecosystem is afflicted by the results of combustion products [86]. Although the combustion of solid combustion items of plastics, ashes, as well as a variety of powders, are produced, which deposited on the floor materials, for example, plants or even soil, they can be found in the aquatic environment. Rainfall results in blurred toxic chemicals tapping in the soil, exerting additional harming outcomes that enable these types of materials to turn out to be incorporated into the food chain. Some of the pollutants chemically interact with water and consequently causing compounds change the pH as well as thus alter the operation of aquatic ecosystems. Nonsoluble compounds introduced to the food chain possess their hazardous consequences. However, incineration decreases the necessity for landfill of plastics waste; however, we can find concerns that hazardous chemicals can be released into the atmosphere in the approach [1]. For that reason basically for this recognized pollution risk, incineration of plastic is much less popular as compared to landfill and mechanical recycling as a waste-management approach. Incineration can be utilized for recovery of an amount of the energy content when we look at the plastic. The valuable energy recovered can differ considerably dependent on if it is utilized for electricity generation, together with heat and power, or as solid refuse fuel for cofuelling of blast furnaces or concrete kilns. Liquefaction to diesel gasoline fuel through pyrolysis can be obtainable and fascination with this process to create diesel fuel is growing most probably owing to increasing oil prices. Energy-recovery processes will be the most appropriate means for managing with extremely mixed plastic such as various electronic and electrical wastes and automotive shredder residual [87–89].

3.2.5 Landfilling of Polyethylene Terephthalate Waste

Landfill was a traditional strategy to waste management; however place for landfill has been starting to become problematic in certain countries. A better-handled landfill site leads to restricted direct environmental problems more than the effects of collection and transport, while we can find prolonged-term issues of pollution of soils and groundwater by many ingredients and breakdown by-products in plastics, which could become chronic organic pollutants. An important problem to landfills from a sustainability point is the fact that none of the landfill material made use of to create the plastic is recovered [90,91]. The disposal of plastics in landfills increases certain issues because the material degrades extremely gradually. Waste materials, for example, paper as well as food wastes are similarly extremely slow to degrade in landfills. Consequently, the degradation of plastics as well as other wastes must not be produced a problem because it possesses small impact on landfill capacity. Plastic wastes do not ever generate problems in landfill functions and do not lead to the toxins of leachate from the landfills. Biodegradation and photodegradation are the couple of solutions which are being investigated for industrial applications. Light as well as air has to be provides for these types of materials to degrade, together with adequate moisture and also nutrients to maintain the microbial activity [92]. These types of demands are not satisfied for materials placed in landfills. In accordance, the EPA does not convince that the utilization of degradable plastics can help eliminate landfill capacity issues. Moreover, the environmental consequence of degradable residues remains poorly understood. Additionally, producing plastics degradable will reduce the quality, as well as overall performance of the material thereby, will not provide some of its primary suitable abilities in several applications [93].

3.3 Conclusion

Plastics are affordable, lightweight and sturdy materials that could easily be molded into a number of products which look for used in an array of applications. An important part of plastic generated every year is employed which will make components of packaging or any other short-lived products which are discarded within per year of manufacture. Bottle-to-fiber recycling offers environmental impact reduction. The understanding explanations show that the reprocessed PET content in a reprocessed bottle is certainly not important into the as a whole environmental performance. Plastic packaging is especially visible as container due to the light in weight character of both flexible and rigid plastics. In the order of reducing environmental desirability: reduce, reuse, and recycle (materials) and recover (energy), while landfill offered least desired control approach. Nowadays, bottle grade PET is the most relevant packaging plastics, and it is probably the most widely used plastics. Almost all of the r-PET flakes were changed into fibers and PET bottles are ubiquitous in our day-to-day lives. These r-PETs can be utilized in a wide range of applications.

References

[1] K. Pivnenko, M.K. Eriksen, J.A. Martín-Fernández, et al., Recycling of plastic waste: presence of phthalates in plastics from households and industry, Waste Manag. 54 (2016) 44–52. Available from: https://doi.org/10.1016/j.wasman.2016.05.014.

[2] R.C. Thompson, C.J. Moore, F.S. vom Saal, S.H. Swan, Plastics, the environment and human health: current consensus and future trends, Philos. Trans. R Soc. B Biol. Sci. 364 (2009) 2153–2166. Available from: https://doi.org/10.1098/rstb.2009.0053.
[3] T. Iwata, Biodegradable and bio-based polymers: future prospects of eco-friendly plastics, Angew. Chem. Int. Ed. 54 (2015) 3210–3215.
[4] I. Hussain, H. Hamid, Plastics in agriculture, Plast. Environ. (2005) 185–209. Available from: https://doi.org/10.1080/02508068908692112.
[5] R.K. Mishra, H.J. Maria, K. Joseph, S. Thomas, Basic structural and properties relationship of recyclable microfibrillar composite materials from immiscible plastics blends: an introduction, Micro and Nanofibrillar Composites (MFCs and NFCs) from Polymer Blends, Elsevier, 2017, pp. 1–25.
[6] R. Kumar Mishra, S. Thomas, N. Karikal, Micro and Nanofibrillar Composites (MFCs and NFCs) from Polymer Blends, Elsevier, 2017.
[7] C. Silvestre, D. Duraccio, S. Cimmino, Food packaging based on polymer nanomaterials, Prog. Polym. Sci. 36 (2011) 1766–1782. Available from: https://doi.org/10.1016/j.progpolymsci.2011.02.003.
[8] L. Chen, R.E.O. Pelton, T.M. Smith, Comparative life cycle assessment of fossil and bio-based polyethylene terephthalate (PET) bottles, J. Clean Prod. 137 (2016) 667–676. Available from: https://doi.org/10.1016/j.jclepro.2016.07.094.
[9] F. Welle, Is PET bottle-to-bottle recycling safe? Evaluation of post-consumer recycling processes according to the EFSA guidelines, Resour. Conserv. Recycl. 73 (2013) 41–45. Available from: https://doi.org/10.1016/j.resconrec.2013.01.012.
[10] D. Amienyo, H. Gujba, H. Stichnothe, A. Azapagic, Life cycle environmental impacts of carbonated soft drinks, Int. J. Life Cycle Assess 18 (2013) 77–92. Available from: https://doi.org/10.1007/s11367-012-0459-y.
[11] A.M.H. Mansour, S.A. Ali, Reusing waste plastic bottles as an alternative sustainable building material, Energy Sustain. Dev. 24 (2015) 79–85. Available from: https://doi.org/10.1016/j.esd.2014.11.001.
[12] Y. Zhu, C. Romain, C.K. Williams, Sustainable polymers from renewable resources, Nature 540 (2016) 354–362. Available from: https://doi.org/10.1038/nature21001.
[13] C.S. Poon, A.T.W. Yu, L.H. Ng, On-site sorting of construction and demolition waste in Hong Kong, Resour. Conserv. Recycl. 32 (2001) 157–172. Available from: https://doi.org/10.1016/S0921-3449(01)00052-0.
[14] F. Convery, S. McDonnell, S. Ferreira, The most popular tax in Europe? Lessons from the Irish plastic bags levy, Environ Resour. Econ. 38 (2007) 1–11. Available from: https://doi.org/10.1007/s10640-006-9059-2.
[15] M.A. Sadi Kazerooni, A. Abdullah, M. Masoud, Navazandeh N. Sajoudi, M.F.M. Kamal, F. Torshizi, R. Taherkhani, Reduce, reuse, recycle and recovery in sustainable construction waste management, Advanced Adv.

Materials Mater. ResearchRes., vol. 446, Trans Tech Publications, 2012, pp. 937–944. Available from: https://doi.org/10.4028/www.scientific.net/AMR.446-449.937.

[16] S.T. Tan, W.S. Ho, H. Hashim, et al., Energy, economic and environmental (3E) analysis of waste-to-energy (WTE) strategies for municipal solid waste (MSW) management in Malaysia, Energy Convers. Manag. 102 (2015) 111–120. Available from: https://doi.org/10.1016/j.enconman.2015.02.010.

[17] L. Shen, E. Worrell, M.K. Patel, Open-loop recycling: a LCA case study of PET bottle-to-fibre recycling, Resour. Conserv. Recycl. 55 (2010) 34–52. Available from: https://doi.org/10.1016/j.resconrec.2010.06.014.

[18] B. Simon, A.M. Ben, R. Földényi, Life cycle impact assessment of beverage packaging systems: focus on the collection of post-consumer bottles, J. Clean Prod. 112 (2016) 238–248. Available from: https://doi.org/10.1016/j.jclepro.2015.06.008.

[19] N. Matar, M.Y. Jaber, C. Searcy, A reverse logistics inventory model for plastic bottles, Int. J. Logist Manag. 25 (2014) 315–333. Available from: https://doi.org/10.1108/IJLM-12-2012-0138.

[20] J. Hopewell, R. Dvorak, E. Kosior, Plastics recycling: challenges and opportunities, Philos. Trans. R Soc. B Biol. Sci. 364 (2009) 2115–2126. Available from: https://doi.org/10.1098/rstb.2008.0311.

[21] J.T. Kirkeby, H. Birgisdottir, T.L. Hansen, et al., Environmental assessment of solid waste systems and technologies: EASEWASTE, Waste Manag. Res. 24 (2006) 3–15. Available from: https://doi.org/10.1177/0734242X06062580.

[22] C. Chan-Halbrendt, D. Fang, F. Yang, Trade-offs between shopping bags made of non-degradable plastics and other materials, using latent class analysis: the case of Tianjin, China, Int. Food Agribus. Manag. Rev. 12 (2009) 179–198.

[23] J.I.M. de Groot, W. Abrahamse, K. Jones, Persuasive normative messages: the influence of injunctive and personal norms on using free plastic bags, Sustain 5 (2013) 1829–1844. Available from: https://doi.org/10.3390/su5051829.

[24] N. Saikia, J. De Brito, Use of plastic waste as aggregate in cement mortar and concrete preparation: a review, Constr. Build. Mater. 34 (2012) 385–401.

[25] T. Chilton, S. Burnley, S. Nesaratnam, A life cycle assessment of the closed-loop recycling and thermal recovery of post-consumer PET, Resour. Conserv. Recycl. 54 (2010) 1241–1249. Available from: https://doi.org/10.1016/j.resconrec.2010.04.002.

[26] M.B. Dangi, C.R. Pretz, M.A. Urynowicz, et al., Municipal solid waste generation in Kathmandu, Nepal, J. Environ. Manag. 92 (2011) 240–249. Available from: https://doi.org/10.1016/j.jenvman.2010.09.005.

[27] M. Abbasi, M.M. Salarirad, I. Ghasemi, Selective separation of PVC from PET/PVC mixture using floatation by tannic acid depressant, Iran. Polym. J. (English Ed) 19 (2010) 483–489.

[28] Y.C. Moh, L. Abd Manaf, Overview of household solid waste recycling policy status and challenges in Malaysia, Resour. Conserv. Recycl. 82 (2014) 50–61.
[29] M. Wegelin, S. Canonica, C. Alder, et al., Does sunlight change the material and content of polyethylene terephthalate (PET) bottles? J. Water Supply Res. Technol.—AQUA 50 (2001) 125–133.
[30] C. Bach, X. Dauchy, M. Chagnon, et al., Chemical migration in drinking water stored in polyethylene terephthalate (PET) bottles: a source of controversy, Water Res. Elsevier 46 (2012) 571–583.
[31] L. Dahlen, A. Lagerkvist, Evaluation of recycling programmes in household waste collection systems, Waste. Manag. Res. 28 (2010) 577–586. Available from: https://doi.org/10.1177/0734242X09341193.
[32] C. Cimpan, M. Rothmann, L. Hamelin, H. Wenzel, Towards increased recycling of household waste: documenting cascading effects and material efficiency of commingled recyclables and biowaste collection, J. Environ. Manag. 157 (2015) 69–83. Available from: https://doi.org/10.1016/j.jenvman.2015.04.008.
[33] R. Woodard, M. Bench, M.K. Harder, The development of a UK kerbside scheme using known practice, J. Environ. Manag. 75 (2005) 115–127. Available from: https://doi.org/10.1016/j.jenvman.2004.11.011.
[34] E. Dimitrakakis, A. Janz, B. Bilitewski, E. Gidarakos, Small WEEE: determining recyclables and hazardous substances in plastics, J. Hazard. Mater. 161 (2009) 913–919. Available from: https://doi.org/10.1016/j.jhazmat.2008.04.054.
[35] B. Ruj, V. Pandey, P. Jash, V.K. Srivastava, Sorting of plastic waste for effective recycling, Int. J. Appl. Sci. Eng. Res. 4 (2015) 564–571. Available from: https://doi.org/10.6088/ijaser.04058.
[36] B. Kuczenski, R. Geyer, Material flow analysis of polyethylene terephthalate in the US, 1996–2007, Resour. Conserv. Recycl. 54 (2010) 1161–1169. Available from: https://doi.org/10.1016/j.resconrec.2010.03.013.
[37] M. Sadat-Shojai, G.R. Bakhshandeh, Recycling of PVC wastes, Polym. Degrad. Stab. 96 (2011) 404–415.
[38] D. Barsky, R. Sala, L. Menéndez, I. Toro-Moyano, Use and re-use: re-knapped flakes from the Mode 1 site of Fuente Nueva 3 (Orce, Andalucía, Spain), Quat. Int. 361 (2015) 21–33. Available from: https://doi.org/10.1016/j.quaint.2014.01.048.
[39] M. Zhang, B. Gao, Removal of arsenic, methylene blue, and phosphate by biochar/AlOOH nanocomposite, Chem. Eng. J. 226 (2013) 286–292. Available from: https://doi.org/10.1016/j.cej.2013.04.077.
[40] Chris C.M.J. Williams, Tapas T. Biswas, Commercial Potential of Giant Reed for Pulp, Paper and Biofuel Production, Rural Industries Research and Development Corporation, 2010.
[41] G. Wu, J. Li, Z. Xu, Triboelectrostatic separation for granular plastic waste recycling: a review, Waste Manag. 33 (2013) 585–597. Available from: https://doi.org/10.1016/j.wasman.2012.10.014.

[42] Garamszeghy, M. Compaction processes and technology for treatment and conditioning of radioactive waste. In: Handbook of Advanced Radioactive Waste Conditioning Technologies, pp. 19-42. 2011.
[43] D.J. Bowman, R.A. Bearman, Coarse waste rejection through size based separation, Miner. Eng. 62 (2014) 102–110. Available from: https://doi.org/10.1016/j.mineng.2013.12.018.
[44] F. Burat, A. Güney, M. Olgaç Kangal, Selective separation of virgin and post-consumer polymers (PET and PVC) by flotation method, Waste Manag. 29 (2009) 1807–1813. Available from: https://doi.org/10.1016/j.wasman.2008.12.018.
[45] G.A. Marques, J.A.S. Tenório, Use of froth flotation to separate PVC/PET mixtures, Waste Manag. 20 (2000) 265–269. Available from: https://doi.org/10.1016/S0956-053X(99)00333-5.
[46] S. Saisinchai, Separation of PVC from PET/PVC mixtures using flotation by calcium lignosulfonate depressant, Eng. J. 18 (2014) 45–53. Available from: https://doi.org/10.4186/ej.2014.18.1.45.
[47] Y. Huang, E. Sutter, N.N. Shi, et al., Reliable exfoliation of large-area high-quality flakes of graphene and other two-dimensional materials, ACS Nano 9 (2015) 10612–10620. Available from: https://doi.org/10.1021/acsnano.5b04258.
[48] H.F. Wehrl, M.S. Judenhofer, S. Wiehr, B.J. Pichler, Pre-clinical PET/MR: technological advances and new perspectives in biomedical research, Eur. J. Nucl. Med. Mol. Imaging 36 (2009) 56–68.
[49] M.I. Eren, S.J. Lycett, Why Levallois? A morphometric comparison of experimental "preferential" Levallois flakes versus debitage flakes, PLoS One 7 (2012). Available from: https://doi.org/10.1371/journal.pone.0029273.
[50] E. Maris, A. Aoussat, E. Naffrechoux, D. Froelich, Polymer tracer detection systems with UV fluorescence spectrometry to improve product recyclability, Miner. Eng. 29 (2012) 77–88. Available from: https://doi.org/10.1016/j.mineng.2011.09.016.
[51] F.B. Dias, L. Plomp, J.B.J. Veldhuis, Trends in polymer electrolytes for secondary lithium batteries, J. Power Sources 88 (2000) 169–191. Available from: https://doi.org/10.1016/S0378-7753(99)00529-7.
[52] D. Patra, A.K. Mishra, Recent developments in multi-component synchronous fluorescence scan analysis, TrAC—Trends Anal. Chem. 21 (2002) 787–798. Available from: https://doi.org/10.1016/S0165-9936(02)01201-3.
[53] V. Sinha, M.R. Patel, J.V. Patel, Pet waste management by chemical recycling: a review, J. Polym. Environ. 18 (2010) 8–25.
[54] A.A. Shah, F. Hasan, A. Hameed, S. Ahmed, Biological degradation of plastics: a comprehensive review, Biotechnol. Adv. 26 (2008) 246–265.
[55] B.A. Miller-Chou, J.L. Koenig, A review of polymer dissolution, Prog. Polym. Sci. 28 (2003) 1223–1270.

[56] A. Göpferich, Mechanisms of polymer degradation and erosion, Biomaterials. 17 (1996) 103–114. Available from: https://doi.org/10.1016/0142-9612(96)85755-3.
[57] J. Morris, Recycling versus incineration: an energy conservation analysis, J. Hazard. Mater. 47 (1996) 277–293. Available from: https://doi.org/10.1016/0304-3894(95)00116-6.
[58] N.J. Schenk, H.C. Moll, J. Potting, The nonlinear relationship between paper recycling and primary pulp requirements, J. Ind. Ecol. 8 (2004) 141–162. Available from: https://doi.org/10.1162/1088198042442379.
[59] A. Buekens, J. Yang, Recycling of WEEE plastics: a review, J. Mater. Cycles Waste Manag. 16 (2014) 415–434.
[60] Vannessa V. Goodship (Ed.), Management, Recycling and Reuse of Waste Composites, Elsevier, Amsterdam, 2009.
[61] L. Rigamonti, M. Grosso, M.C. Sunseri, Influence of assumptions about selection and recycling efficiencies on the LCA of integrated waste management systems, Int. J. Life Cycle Assess 14 (2009) 411–419. Available from: https://doi.org/10.1007/s11367-009-0095-3.
[62] A. Damgaard, A.W. Larsen, T.H. Christensen, Recycling of metals: accounting of greenhouse gases and global warming contributions, Waste Manag. Res. 27 (2009) 773–780.
[63] A.C. Woolridge, G.D. Ward, P.S. Phillips, et al., Life cycle assessment for reuse/recycling of donated waste textiles compared to use of virgin material: an UK energy saving perspective, Resour. Conserv. Recycl. 46 (2006) 94–103. Available from: https://doi.org/10.1016/j.resconrec.2005.06.006.
[64] F. Perugini, M.L. Mastellone, U. Arena, A life cycle assessment of mechanical and feedstock recycling options for management of plastic packaging wastes, Environ. Prog. 24 (2005) 137–154.
[65] G.P. Karayannidis, D.S. Achilias, Chemical recycling of poly(ethylene terephthalate), Macromol. Mater. Eng. 292 (2007) 128–146. Available from: https://doi.org/10.1002/mame.200600341.
[66] Q. Wang, X. Yao, S. Tang, et al., Urea as an efficient and reusable catalyst for the glycolysis of poly(ethylene terephthalate) wastes and the role of hydrogen bond in this process, Green Chem. 14 (2012) 2559. Available from: https://doi.org/10.1039/c2gc35696a.
[67] M.I. Jahirul, M.G. Rasul, A.A. Chowdhury, N. Ashwath, Biofuels production through biomass pyrolysis—a technological review, Energies 5 (2012) 4952–5001.
[68] M. Genta, F. Yano, Y. Kondo, et al., Development of chemical recycling process for post-consumer PET bottles by methanolysis in supercritical methanol, Mitsubishi Heavy Ind. Ltd. Tech. Rev. 40 (2003) 1–4.
[69] B. Zamani, M. Svanström, G. Peters, T. Rydberg, A carbon footprint of textile recycling: a case study in Sweden, J. Ind. Ecol. 19 (2015) 676–687. Available from: https://doi.org/10.1111/jiec.12208.

[70] L. Shen, E. Worrell, M.K. Patel, Comparing life cycle energy and GHG emissions of bio-based PET, recycled PET, PLA, and man-made cellulosics, Biofuels Bioprod. Biorefining 6 (2012) 625–639. Available from: https://doi.org/10.1002/bbb.1368.
[71] F. Intini, S. Kühtz, Recycling in buildings: an LCA case study of a thermal insulation panel made of polyester fiber, recycled from post-consumer PET bottles, Int. J. Life Cycle Assess 16 (2011) 306–315. Available from: https://doi.org/10.1007/s11367-011-0267-9.
[72] L. Shen, E. Nieuwlaar, E. Worrell, M.K. Patel, Life cycle energy and GHG emissions of PET recycling: change-oriented effects, Int. J. Life Cycle Assess 16 (2011) 522–536. Available from: https://doi.org/10.1007/s11367-011-0296-4.
[73] L. Shen, E. Worrell, M.K. Patel, Environmental impact assessment of man-made cellulose fibres, Resour. Conserv. Recycl. 55 (2010) 260–274. Available from: https://doi.org/10.1016/j.resconrec.2010.10.001.
[74] Patagonia Inc., Patagonia's Common Threads Garment Recycling Program : A Detailed Analysis, Patagon Inc., 2005, pp. 1–11. pp.
[75] H. Yuan, Z. He, Integrating membrane filtration into bioelectrochemical systems as next generation energy-efficient wastewater treatment technologies for water reclamation: a review, Bioresource Bioresour. TechnologyTechnol. 195 (2015) 202–209. Available from: https://doi.org/10.1016/j.biortech.2015.05.058.
[76] D.E. Nikles, M.S. Farahat, New motivation for the depolymerization products derived from poly(ethylene terephthalate) (PET) waste: a review, Macromol. Mater. Eng. 290 (2005) 13–30.
[77] K. Ramadevi, R. Manju, experimental investigation on the properties of concrete with plastic PET (bottle) fibres as fine aggregates, J. Emerg. Technol. Adv. Eng. 2 (2012) 42–46.
[78] K. Kohlgrüber (Ed.), Co-Rotating Twin-Screw Extruder, Carl Hanser Verlag GmbH Co KG, Verlag, 2012.
[79] R.K. Mishra, A. Shaji, S. Thomas, Processing, rheology, barrier properties, and theoretical study of microfibrillar and nanofibrillar in situ composites, Micro and Nano Fibrillar Composites (MFCs and NFCs) from Polymer Blends, Elsevier, 2017, pp. 233–262.
[80] O.C. Aja, H.H. Al-Kayiem, Review of municipal solid waste management options in Malaysia, with an emphasis on sustainable waste-to-energy options, J. Mater. Cycles Waste Manag. 16 (2014) 693–710.
[81] O. Eriksson, G. Finnveden, Plastic waste as a fuel-CO 2-neutral or not? Energy & Environmental. ScienceSci. 2 (9) (2009) 907–914. Available from: https://doi.org/10.1039/b908135f.
[82] S.L. Wong, N. Ngadi, T.A.T. Abdullah, I.M. Inuwa, Current state and future prospects of plastic waste as source of fuel: a review, Renew. Sustain. Energy Rev. 50 (2015) 1167–1180.

[83] A. Williams, J.M. Jones, L. Ma, M. Pourkashanian, Pollutants from the combustion of solid biomass fuels, Prog. Energy Combust. Sci. 38 (2012) 113–137.
[84] B.R.T. Simoneit, Biomass burning—a review of organic tracers for smoke from incomplete combustion, Appl. Geochem. 17 (2002) 129–162.
[85] D. Wang, A. Bao, W. Kunc, W. Liss, Coal power plant flue gas waste heat and water recovery, Appl. Energy 91 (2012) 341–348. Available from: https://doi.org/10.1016/j.apenergy.2011.10.003.
[86] J. Prapas, M.E. Baumgardner, A.J. Marchese, et al., Influence of chimneys on combustion characteristics of buoyantly driven biomass stoves, Energy Sustain. Dev. 23 (2014) 286–293. Available from: https://doi.org/10.1016/j.esd.2014.08.007.
[87] A. Bosmans, I. Vanderreydt, D. Geysen, L. Helsen, The crucial role of waste-to-energy technologies in enhanced landfill mining: a technology review, J. Clean Prod. 55 (2013) 10–23. Available from: https://doi.org/10.1016/j.jclepro.2012.05.032.
[88] L. Zheng, J. Song, C. Li, et al., Preferential policies promote municipal solid waste (MSW) to energy in China: current status and prospects, Renew. Sustain. Energy Rev. 36 (2014) 135–148.
[89] S.L. Low, Ma Connor, G.H. Covey, Turning mixed plastic wastes into a useable liquid fuel, 6th World Congr. Chem. Eng. (2001) 6. p.
[90] N. Marchettini, R. Ridolfi, M. Rustici, An environmental analysis for comparing waste management options and strategies, Waste Manag. 27 (2007) 562–571. Available from: https://doi.org/10.1016/j.wasman.2006.04.007.
[91] F. Cherubini, S. Bargigli, S. Ulgiati, Life cycle assessment (LCA) of waste management strategies: landfilling, sorting plant and incineration, Energy 34 (2009) 2116–2123. Available from: https://doi.org/10.1016/j.energy.2008.08.023.
[92] O. Dada, C. Mbohwa, Biogas upgrade to biomethane from landfill wastes: a review, Procedia Manuf. 7 (2017) 333–338. Available from: https://doi.org/10.1016/j.promfg.2016.12.082.
[93] N.C. Billingham, M. Bonora, D. De, Environmentally degradable plastics based on oxo-biodegradation of conventional polyolefins, Biodegrad. Polym. Plast. 44 (2003) 313–325. Available from: https://doi.org/10.1007/978-1-4419-9240-6_21.

4 Chemical Depolymerization of PET Bottles via Glycolysis

Anvita Sheel and Deepak Pant
Department of Environmental Sciences, Central University of Himachal Pradesh, Dharamshala, Himachal Pradesh, India

4.1 Introduction

Recycling of polyethylene terephthalate (PET) (Fig. 4.1) appears to be the most attractive method for the associated waste management because it provides opportunities for reducing the use of raw petrochemical products to minimize carbon dioxide emissions and decrease the load on landfill space. In 2008, worldwide production capacity of PET was estimated to be to 64,400 kt per year [1] and is continuously increasing. It has been expected that the global PET packaging consumption will grow to over 20 Mt by 2019 [2].

Existing PET recycling methods (Scheme 4.1) can be summarized in four categories: (i) primary, (ii) secondary, (iii) tertiary, and (iv) quaternary. There is also a parallel concept of "zero-order" recycling which involves the direct reuse of a PET waste material [3]. Here the plastic articles remain undamaged and are reused in their original form, like as refilling a mineral water bottle.

4.1.1 Primary Recycling

This oldest method of recycling is also known as reextrusion and implies the "in-plant" recycling of uncontaminated, clean industrial scrap materials. It is a simple, low-cost process, suitable for single type of uncontaminated scrap and not a popular choice of recyclers [4,5].

4.1.2 Secondary Recycling

It is a mechanical recycling or physical reprocessing technique and involves a series of steps like melting, shredding or granulation of waste plastic. PET must be separated from other waste plastics like PVC,

Recycling of Polyethylene Terephthalate Bottles.
DOI: https://doi.org/10.1016/B978-0-12-811361-5.00004-3

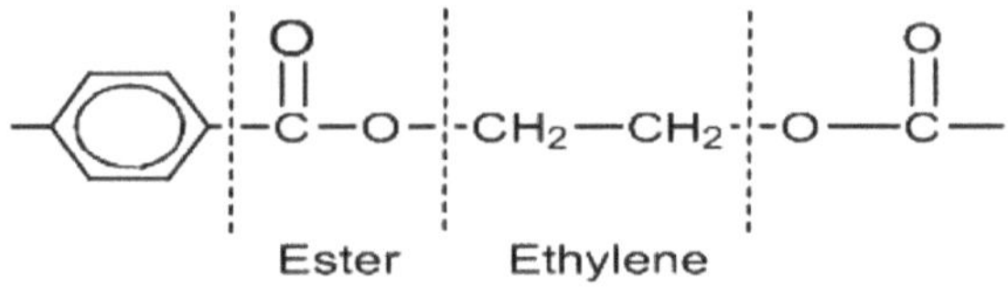

Figure 4.1 Polyethylene terephthalate (PET).

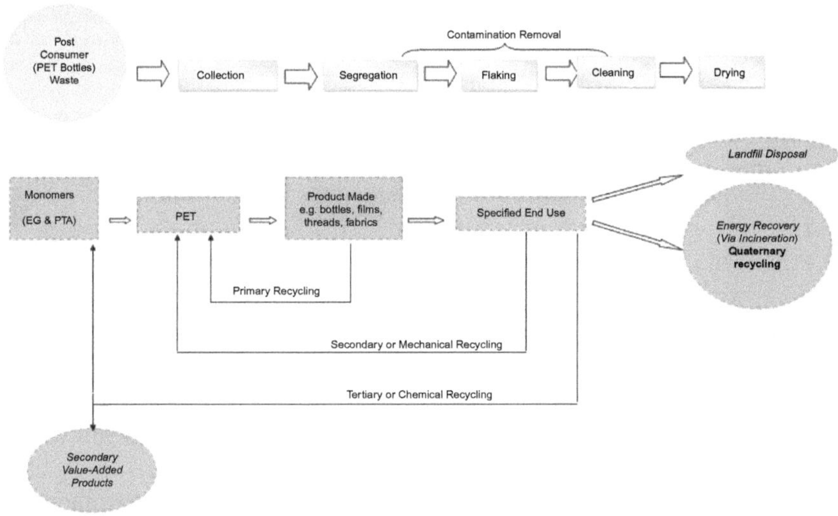

Scheme 4.1 PET recycling.

polyethylene, etc. Extrusion in the presence of contaminants reduces the MW due to degradation [6]. Therefore after grinding, PET flakes are given a wash either with tetrachloroethylene (TCE) or with NaOH (2% solution) and detergent at 80°C followed by a cold water wash to remove the present impurities or contamination. Although physical recycling of PET bottles is relatively simpler and economic it has the main disadvantage due to thermal and hydrolytic degradation reactions responsible for the reduction in melt viscosity. Furthermore, due to water and other trace acidic impurities chain-scission reaction occurs, which reduces the molecular weight of recycled resin. Consequently, there is deterioration of product properties in every cycle. Given that mechanically reprocessed PET has changed product properties such as melt viscosity and molecular weight; there may be a need to add additional antioxidants, processing aids or other additives to recycle resin to produce a plastic with the preferred qualities [7].

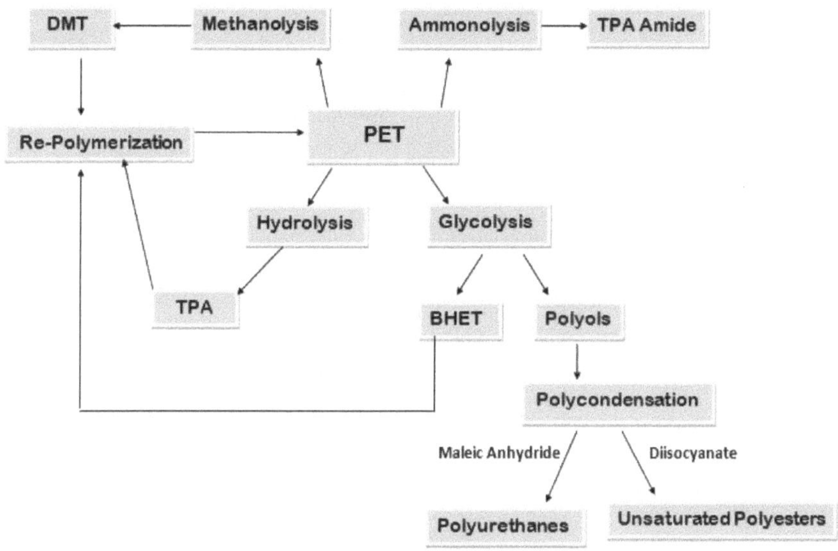

Scheme 4.2 Chemical depolymerization of PET.

4.1.3 Tertiary or Chemical Recycling

Chemical recycling/depolymerization (Scheme 4.2) fundamentally involve the conversion of polymer chain. The polymer backbone is depolymerized into monomer units along with oligomers (randomly raptured larger chain fragments) with related evolution of gaseous products. Chemical recycling is done by solvolysis which means of using various solvents including water to bring about the depolymerization of PET. Following are the various techniques for chemical recycling:

a. *Hydrolysis*

   It involved the depolymerization of post consumed PET bottles into terephthalic acid (TPA) and ethylene glycol (EG) (Figs. 4.2 and 4.3). Hydrolysis of PET can be occurred in alkaline, acid, and natural condition.

b. *Methanolysis*

   It uses methanol at high pressures and temperatures for the degrading PET and give dimethyl terephthalate (DMT) and EG along with oligomers as end products [8] (Fig. 4.4). Although it is currently and effectively applied to scrap bottles, fiber waste, used films, and plant waste [9] high cost of separation and refinement in combination of the reaction products (glycols, alcohols, and phthalate

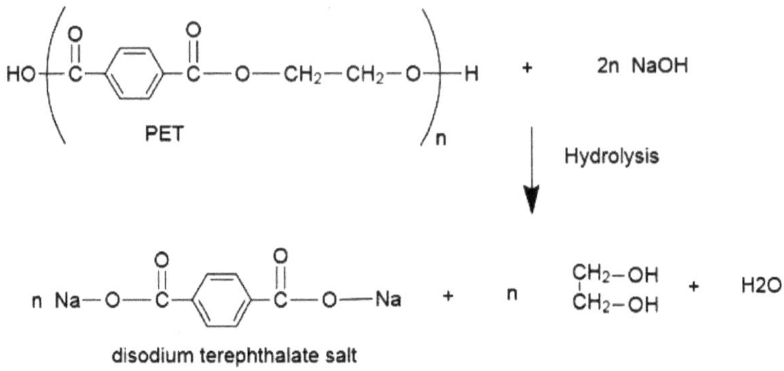

Figure 4.2 Alkaline hydrolysis.

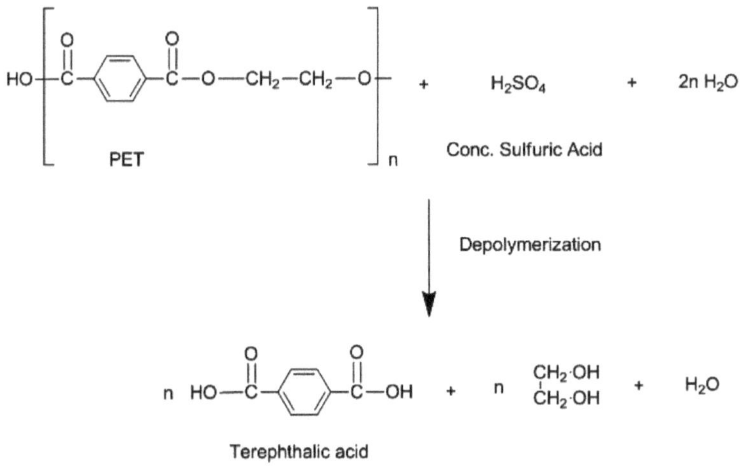

Figure 4.3 Acidic hydrolysis.

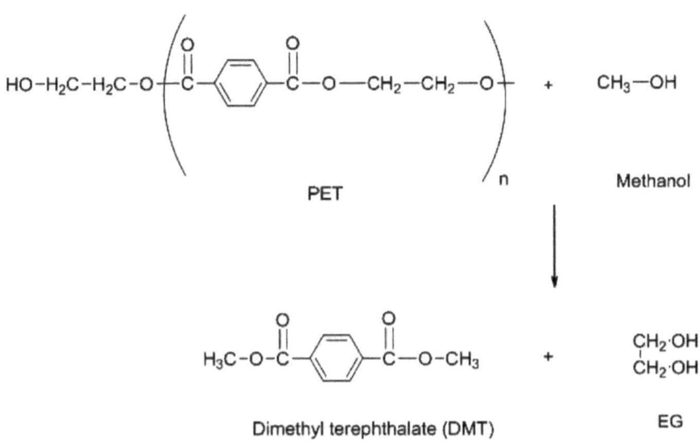

Figure 4.4 Methanolysis.

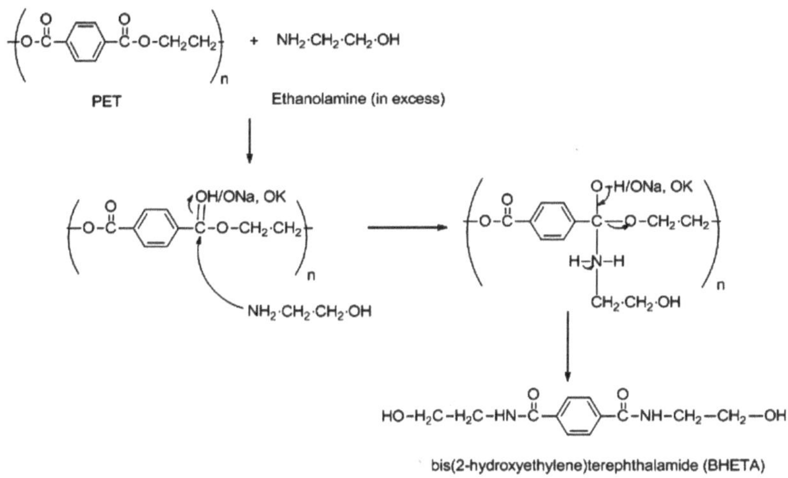

Figure 4.5 Aminolysis [24].

derivatives still remains a drawback to this technique. Water perturbs the process by poisoning the catalyst and forming various azeotropes.

c. *Aminolysis*

The aminolysis of PET produces diamides of TPA like bis (2-hydroxyethylene)terephthalamides (BHETA) (Fig. 4.5). Partial aminolysis improves the properties of PET which is to be used in the manufacture of fibers with certain predefined character [10]. This process has not yet been commercialized.

d. *Ammonolysis*

When ammonia reacts with PET in an EG environment terephthalamide is produced which can further be converted into terephthalic acid nitrile and subsequently to p-xylylenediamine or 1,4-bis(aminoethyl)cyclohexane [11]. Ammonolysis has been successfully applied to postconsumer PET bottles waste (Fig. 4.6).

e. *Glycolysis*

Discussed in Section 2.

4.1.4 Quaternary Recycling

It characterizes recovering energy content from the plastic waste by incineration. When collection, segregation and further recycling of plastic waste are difficult due to economy, toxic or hazardous nature, etc.

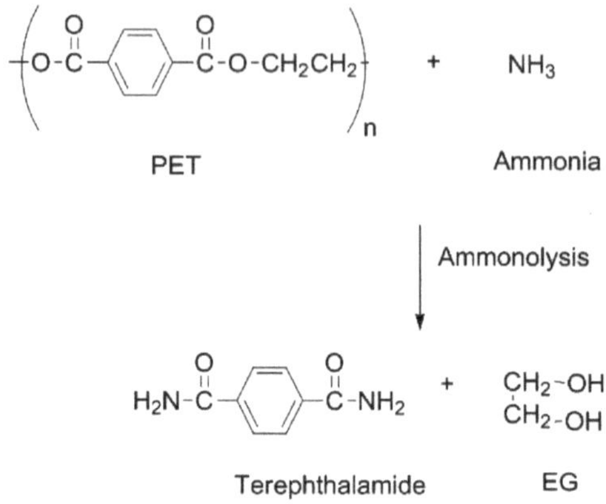

Figure 4.6 Ammonolysis.

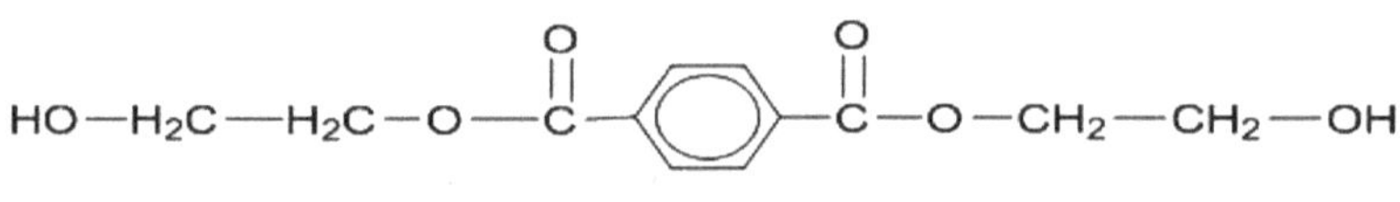

Figure 4.7 Bis(2-hydroxyethyl) terephthalate (BHET).

energy recovery is the best management option. This is usually done through incineration. However, this method has environmental issues as burning plastic releases air borne toxic substances and pose potential health risks for all living beings.

4.2 Glycolysis

Glycolysis is a kind of solvolysis reaction, using glycol (1,2-ethanediol) and/or it is derived as the reactive solvent. Glycolysis is an important and commercial process for chemical depolymerization of PET. Chemically, glycolysis reaction is the "molecular degradation of PET polymer by glycols, in the presence of trans-esterification catalysts, where the ester linkages breaks and replace with hydroxyl terminals" [12].

The main products obtained from EG [13], diethylene glycol [14], propylene glycol, and dipropylene glycol [15] are bis(2-hydroxyethyl) terephthalate (BHET) (Fig. 4.7) and other oligomers (PET glycolyzates).

These products can be used further for manufacturing articles like polyurethane foams, copolyesters, hydrophobic dye stuffs, unsaturated

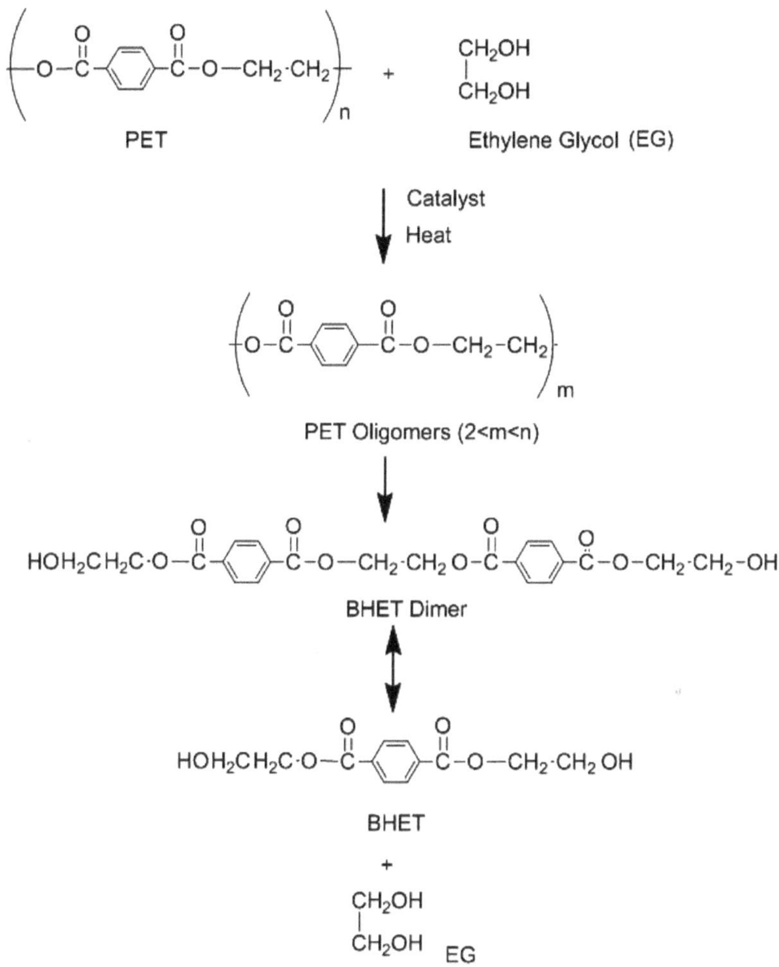

Figure 4.8 Glycolysis.

resins and acrylic coatings. The produced BHET with fresh BHET can be used for PET production using any of the method based on either DMT or TPA. Glycolysis involving following steps (Fig. 4.8):

1. In initiation step, carbonyl carbon of ester group of the polymer is attacked by the free electron pair present on EG.
2. Hydroxyl ethyl group of EG bonds with carbonyl carbon of PET which breaks long polymer chain into short oligomers with subsequent formation of BHET.

There are four key methods of glycolysis techniques:

1. Catalyzed glycolysis
2. Solvent-assisted glycolysis
3. Supercritical glycolysis
4. Microwave-assisted glycolysis

4.2.1 Catalyzed Glycolysis

It is the most promising method for increasing the rate of glycolysis of PET using catalyst (Fig. 4.9 and Table 4.1). Even though a lot of factors like time, temperature, PET/solvent ratio, etc. influences the rate of glycolysis. Kinetic study of the process shows that without catalyst glycolysis of PET is extremely slow and complete depolymerization cannot be obtained. Metal based trans-esterification catalysts are commonly uses to accelerate the rate of reaction.

The transformation of dimer into BHET monomer is a reversible process. Metal based catalysts facilitate the attack of EG on PET and forms complex with carbonyl group, leading to the formation of BHET. Various glycolytic depolymerization processes involving catalysts like:

1. Metal derivatives
2. Zeolites
3. Ionic liquids

4.2.1.1 Metal Derivatives

Metal acetates are the first group of catalysts reported for PET glycolysis. Vaidya and Nadkarni in 1987, used zinc acetate to produce polyester polyols from waste PET. The use of metal acetates (manganese, zinc, lead, and cobalt) as catalyst provides some interesting finding about his activities:

1. Zinc acetate (ZnAc) is the best activity, and out of various metal acetates catalytic activities follow the trend $Zn^{+2} > Mn^{+2} > Co^{+2} > Pb^{+2}$ [25].
2. PET glycolysis using ZnAc at 190°C achieved equilibrium in 02 h [26] and with manganese acetate the time is reduced to 1.5 h [27].

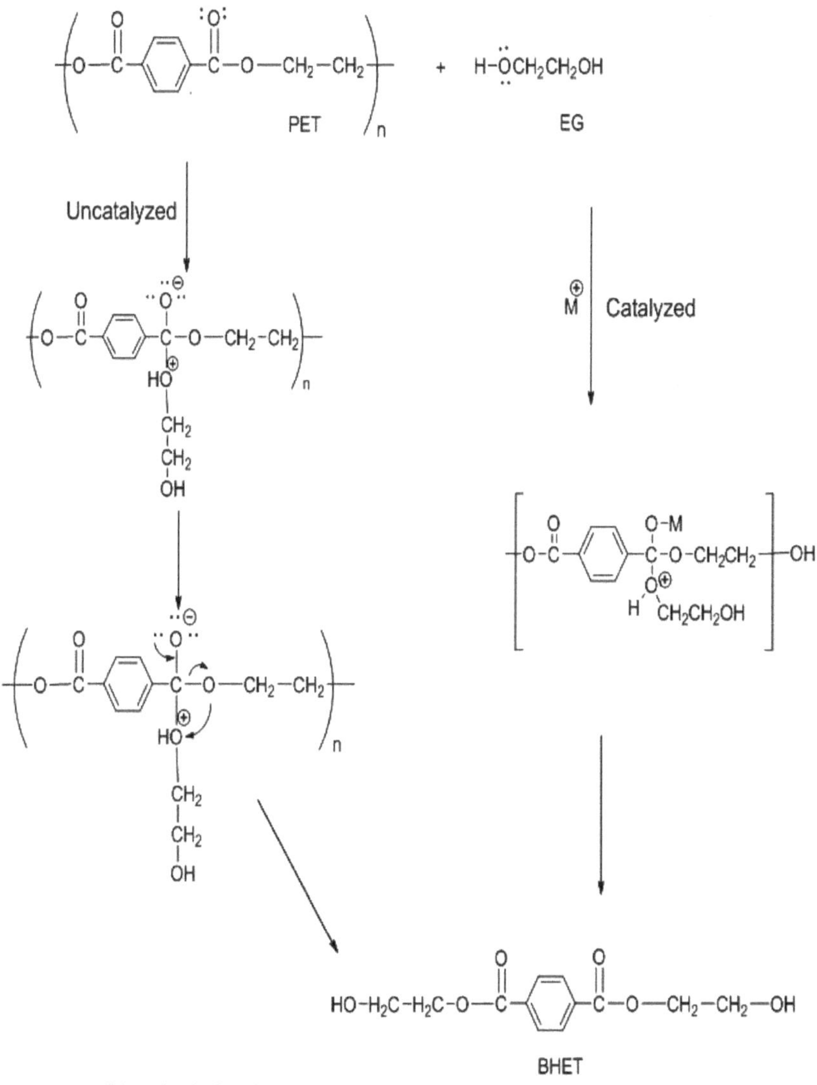

Figure 4.9 Glycolysis in absence and presence of catalyst.

Most favorable and almost complete glycolytic conversion (98.66%) of PET (197°C, 90 min, particle size 127.5 μm) resulting into retrieval of monomeric products like EG and DMT [28]. The optimal PET particle size was obtained by studying its weight loss vs time correlation. Although metal salts effectively enhance PET glycolysis, they cease to advance the reaction above temperature of 245°C owing it to the limitations imposed by mass-transfer restrictions [29].

Table 4.1 Catalytic Glycolysis

Catalyst Used Yield % BHET	Reaction Conditions				References
	Temperature (°C)	Time (min)	EG/PET Ratio	PET/Catalyst Weight Ratio	
Zinc acetate (85.6%)	196	180	5 (w/w)	0.01	[16]
Zinc acetate (62.8% in the product), titanium phosphate (97.5% in the product)	200	150	2.77 (mol/mol)	0.003	[17]
Lead acetate (62.51%), sodium carbonate (61.65%), sodium bicarbonate (61.5%), acetic acid (61.94%)	190	480	6 (mol/mol)	0.005	[18]
Acetic acid (62.42%)	190	480	6 (mol/mol)	0.005	[30]
Lithium hydroxide (63.50%)					
Sodium sulfate (65.72%)					
Potassium sulfate (64.42%)					
Zinc chloride (73.24%)	197	480	10 (mol/mol)	0.005	[19]
Lithium chloride (59.46%)					
Didymium chloride (71.01%)					
Magnesium chloride (55.67%)					
Ferric chloride (56.28%)					
Manganese oxide on silica nanoparticles Mn_3O_4/SNPs (> 90%)	300	80	11:1	1.0	[20]
Zinc oxide on silica nanoparticles—$ZnMn_2O_4$ (92.2%)	260	60	17.2	1.0	[21]
Graphene oxide–manganese oxide—(GO–Mn_3O_4) (96.4%)	300	80	3.6	1.0	[22]
Mn_3O_4 without graphene oxide (82.7%)					
Different ionic liquids	190	120	10 (w/w)	0.05	[34,35]
100% conversions					
[Bmim]OH (71.2%)	190	120	10	0.05	[23]
[Bmim][OAc] (58.2%)	190	180	6.6	3	[24]
$Cu(OAc)_2$–[Bmim][OAc] (53.95%)					
$Zn(OAc)_2$–[Bmim][OAc] (45.6%)					

With the use of titanium (IV) phosphate as the catalyst, glycolysis was achieved at 2.5 h at 200°C (0.003 catalyst: PET, weight ratio). The resultant product contained considerably higher amount of BHET (97.5%) in comparison to the reaction ZnAc (62.8%) [17].

Zinc and lead as heavy metal shows rather well-known detrimental effects on the environment. Less harmful carbonates and bicarbonates of sodium can proposed as valuable environmental solution with comparable monomer yields [18]. Some other alternate in this direction is sodium and potassium sulfate, glacial acetic acid, lithium hydroxide with almost similar monomer yields [30].

4.2.1.2 Zeolites

The catalytic activity of zeolites is due to the presence of active sites and surface area in their meso- and micropores. They perform well up to a temperature below 245°C. However, using zeolites as catalyst for glycolysis of PET did not show any significant improvement in terms of BHET yield and the maximum yield of BHET monomer was around 73%.

For improving the catalytic activity by expanding the surface area of active sites combination of metal oxides and zeolites at nanoscale was proposed. The intrinsic properties of the catalysts may vary at nanoscale, resulting in increased effectiveness as compared to that of their bulk equivalent [31,32]. The metal oxide catalysts impregnated in silica [silica microparticles (SMPs) or silica nanoparticles (SNPs)] was the next modification in this direction [20]. Using manganese, zinc, and cerium oxides deposited on SNPs, BHET yield was 90%. The reaction attained equilibrium in 80 min at 300°C, 1.1 MPa keeping EG:PET ratio 11 and PET:catalyst weight ratio 0.01. Furthermore, smaller size of the support leads to better distribution of the catalysts and elevated chance of contact between support and catalyst due to increased surface-to-volume ratio [33].

4.2.1.3 Ionic Liquids

Ionic liquid (IL) is a salt having melting point lower than 100°C, in general. They are of great interest to scientist because they exhibit certain distinctive characteristics, such as thermal stability, strong solvation for both organic and inorganic compounds, electrochemical stability, nonvolatility and low flammability. IL can potentially replace the traditional organic solvents used for degrading PET [34]. Wang et al.

were the first group to utilize IL catalyst for PET glycolysis. The main benefit of using IL as catalyst is its easy removal from the final products and repetitive utilization. Using 1-butyl-3-methylimidazolium bromide ([Bmim]Br) as catalyst with reaction time of 8 h at 180°C, 100% PET conversion was achieved. These reactions generally follow first order kinetics (activation energy 232.79 kJ/mol). Fe containing magnetic IL shows better catalytic activity than the conventional catalysts [35]. The higher amount of catalyst for ILs can justified due to recover and reusability of these catalysts.

Yue et al. in 2011 modified [Bmim] based ionic liquid, and proposed the higher catalytic activity of basic [Bmim]OH as compared to [Bmim]Br and [Bmim]Cl (reaction conditions: time 2 h, 190°C, EG: PET molar ratio 10, catalyst:PET weight ratio 0.05) and reported 71.2% BHET and 100% PET conversion.

In an industrial set-up, acetate based IL [Bmim][OAc] was a prospective catalyst with reusability up to six times. Furthermore, anion and cation in this Lewis base IL show synergistic effect and aid the attack of oxygen in the ethylene on the carbocation of ester group. Following a first order reaction kinetics in 3 h, it gives 100% PET conversion and 58.2% BHET yield (activation energy 58.53 kJ/mol) [24]. Another investigation in the possible catalyst by the same group of researchers lead to the use of 1-butyl-3-methylimidazolium acetate-promoted zinc acetate ($Zn(OAc)_2$–[Bmim][OAc]) and 1-butyl-3-methylimidazolium acetate-promoted copper acetate ($Cu(OAc)_2$–[Bmim][OAc]) as catalysts. Results reported complete PET conversion and BHET yield of 53.95% and 45.6%, respectively, in the similar reaction conditions and following first order reaction with activation energy of 56.4 kJ/mol and 53.8 kJ/mol, respectively. The reaction mechanism was proposed as:

1. Development of hydrogen bonding between carbonyl oxygen of metal acetate and C_2–H on imidazolium ring.
2. Metal acetate changes bidentate to a unidentate coordination structure and supports the coordination of carbonyl oxygen of PET molecule with the metal in acetate. Facilitating the attack of oxygen atom (hydroxyl group of EG) on the carbon cation of the ester group of PET.

4.2.2 Solvent-Assisted Glycolysis

Solvent-assisted glycolysis involves the depolymerization of PET with EG in the presence of another solvent as a reaction medium. When xylene (C_8H_{10}) was added in PET glycolysis reaction catalyzed by ZnAc; it provided mix ability to PET–glycol mixture and improved the yield of BHET. At 170–225°C, EG dissolves sparingly in xylene but dissolves readily in PET, also glycolyzed products are soluble in xylene. Consequently, reaction move forward and glycolyzed products travel from the PET glycol phase to xylene phase.

Wang and group in 2015 reported the option of using deep eutectic solvents for catalyst in support of mild and fast glycolysis of PET. Such solvents have low toxicity, ease in preparation, low cost, good biological compatibility and their comparable features with ionic liquids. They synthesized a number of urea–metal salt mixtures by combining urea and metal salts at different molar ratios. 100% conversion of PET and 83% selectivity of BHET was reported in 38 min at the optimal conditions of EG (20 g), PET(5 g), catalyst (n(urea)/n($ZnCl_2$): 4/1, 0.25 g) at 170°C under atmospheric pressure. The synergetic catalysis of H-bond and coordination bond formed between eutectic solvent catalyst and EG were the assigned reason for the elevated catalytic activity.

100% conversion of PET waste was recorded when urea was used as a green, low-priced, highly active, easily and extensively produced catalyst for degradation of PET [36]. The ratio of PET waste:EG:urea was 1:4:0.1 and used under atmospheric pressure at 160°C for 2.5 h. This experiment demonstrated that urea worked well and efficiently, even though it was utilized 10 times. Investigators further proposed a mechanism based on density function theory and deduced that formation of H-bonds between oxygen of carbonyl of urea and EG increases the O–H bond length of hydroxyl in EG making loss of H atom easier. This increases in the nucleophilicity of the oxygen in EG. Consequently, this hydroxyl oxygen attacks the carbon of ester group in PET with much more ease and finally encourages the glycolysis of PET. According to this cognition, they further inferred that catalysts able to form H-bonds with alcohols can enhance the degradation of PET.

4.2.3 Supercritical Glycolysis

In supercritical glycolysis, degradation of PET by EG is done at a temperature and pressure above the critical point of EG (446.7°C and 8.2 MPa). Using supercritical fluids in the reaction eradicate the need of using catalyst (which sometimes presents difficulty in separation from reaction products) to make the process environmentally preferable. Reaction was done at 450 °C, 15.3 MPa and obtained results were compared to those of the subcritical processes done at 350, 300 °C with pressure 2.49, 1.1 MPa, respectively [37]. For supercritical method the equilibrium between dimer and BHET was attained early and in 30 min BHET yield reached its maximum 93.5%. This method also suppressing the yield of by-products and provides a very high yield of BHET.

4.2.4 Microwave-Assisted Glycolysis

Heat supply from microwave radiation decreases the reaction time of glycolysis considerably, from 8 h to just 35 min [38] without affecting the yield of BHET monomer. The research is focused on the development of more efficient catalyst to use along with microwaves radiations that might help to augment monomer yield and reduces overall reaction time. Cho et al. [39] studied non-aqueous, microwave irradiated glycolysis in alkaline atmosphere for the development of economically reasonable treatment process with reduced reaction time. The effectiveness of the process depended on two parameters: the alkali catalyst concentration and of course the irradiation time that had an effect on the reaction temperatures. Chaudhary et al. [40] demonstrated that using microwaves can reduce total glycolysis reaction time to approximately 30 min.

4.3 Factors Influencing Glycolysis

4.3.1 Contaminant

Contaminations like as adhesives, coloring agents, etc. present in the postconsumer PET articles responsible for the physical and chemical properties of the product [41]. Moisture left from flakes washing reduces the molecular weight of PET through a hydrolysis reaction [8]. Adhesives (as little as 100 ppm) can cause chain scission reactions [42]. Other contaminants like remains of colored bottles, printed stickers cause undesirable coloring throughout processing. Acetaldehyde which is a by-product of PET degradation reactions presents problems and stabilizers like diphenylamine and 4-aminobenzoic acid are often added to curtail the quantity of acetaldehyde [43].

4.3.2 Stirring Speed

The effect of stirring speed under same reaction conditions was studied by Haghighi et al. [44] and it was concluded that with increasing the stirring speed, reaction efficiency increases, up to a certain point. After that stirring speed has no effect on the reaction. The optimal rate of stirring was estimated at 400–600 rpm. At low speed (50 rpm) very low reaction efficiency and long reaction time was reported [45].

4.3.3 Particle Size

Generally, smaller particle size leads to a better PET decomposition due to higher surface area. Particles with smaller dimensions 0.25–0.35 mm are found suitable for degradation [46].

4.3.4 Solvent Ratio

EG is mostly used for glycolytic depolymerization of PET. The most important factor affecting PET glycolysis is glycol/PET ratio. Depolymerization rate is proportional to the square of EG concentration [27]. By increasing ratio, the rate of decomposition increases and leads to producing lower MW oligomers. To increase the mixability xylene can often been used.

4.3.5 Reactions Conditions

PET glycolysis has been studied over a wide range of reaction conditions like temperature, time, pressures, type, and amount of catalyst used. Effect of different reaction parameters follows the trends: glycolysis time $<$ glycolysis temperature $<$ catalyst concentration [47]. Optimum reaction conditions and product yield can vary in different conditions (refer to Table 4.1). At prolonged reaction time in presence of excess quantities of EG, BHET can polymerize into dimer or higher oligomers.

4.4 Advantages and Drawbacks of Glycolysis

Advantages

- The process can be incorporated into fresh PET production plants. Recovered BHET (after purification) can be mixed with fresh BHET.

- It is commercially applied for medium and smaller installations and is economically viable in terms of cost and application.
- PET glycolyzates find numerous applications for example in the fabrication of unsaturated polyesters, polyurethane foams, polyisocyanurate foams, urethane oils, etc.
- PET glycolysis can be selected for lower chain length oligomers.
- Ionic liquids and microwave assisted technique makes the process green, further more ionic liquid makes the purification of products simpler.
- Microwave assisted technique significantly reduces the reaction time.
- Using supercritical fluids eliminates the use of catalyst (and associated separation issues) altogether.

Drawbacks

- Progression of reaction is too slow without catalyst.
- Conventional techniques sometimes do not adequately purify reaction products.
- Most of the catalysts used are soluble in EG, thus requiring an additional unit for its separation.
- Nanocomposites and organic solvents have their associated harms to the environment.

4.5 Analytical Techniques

The employment of glycolyzed products should have properties comparable to their analogs made from using virgin materials. Various instruments provide the technological and theoretical means to gain a better understanding of the material properties needed to design and optimize polymeric materials for further applications. Some of these are:

1. Nuclear magnetic resonance spectroscopy (C-NMR and H-NMR), Fourier transform infrared spectroscopy, and Raman spectroscopy can be used for structural elucidation of prepared polymeric materials.

2. Gel permeation chromatography (GPC)—determines the molecular weight distribution of polymers within detectable range of molecular weights from a few thousand to hundreds of thousands of Dalton.
3. X-ray diffraction (XRD)—for quantitative appraisal of crystallinity, crystallite size and changes in crystalline phases of polymers; for potentially identifying semicrystalline polymers based on their crystalline structures.
4. Differential scanning calorimetry (DSC)—to study glass or melt transition temperatures, crystallinity and the thermal and processing history of polymer materials.
5. Thermogravimetric analysis (TGA)—is a technique that helps in assessing the thermal stability of polymers. Thermogravimetric curves also permit us to make out about phase segregation in polymers.
6. Tensile and impact tester assists determine mechanical features of polymers such as tensile strength, tear strength, modulus, elongation at break, etc.

4.6 Applications of the Glycolyzed Products

Glycolysis offers a green option for its management. To improve the economic viability of the process efforts have been made to synthesize value added products from its depolymerization, some of these are as follow:

4.6.1 Polyurethane Products

The glycolyzed oligoesters produced can be made to react with aliphatic diacids to obtain polyester polyols. These polyols serve as raw material in polyurethane manufacturing. They can also be directly reacted with diisocyanate compounds to fabricate a urethane group; that can further be modified to form flexible or rigid polyurethane foams. Some of these rigid foams obtained with degrading PET using triethanolamine can be regarded as insulation materials [48]. Ristic et al. [49] used recycled products of PET, oligoesters for anionic polyurethane dispersion preparation with isophorone diisocyanate (IPDI), dimethylolpropionic acid (DMPA), and ethylene dimine (EDA). Oligoester diols can also be utilized as raw materials for adhesive compositions such as epoxy resins to be used as various surface (i.e., wood, metal) coatings.

Cross-linked polyurethane obtained by combining with high quantities of aliphatic isocyanates after curing showed good mechanical properties [50]. Akar et al. [51] developed urethane oil varnishes by first reacting (trans-esterification reaction) the glycolyzed PET waste with glycolyzed products of soy bean oil and then subsequently treating this reaction mixture with toluene diisocyanate. Varnish oils thus produced exhibited broad spectrum physical properties, some characteristics were even better than those of commercial grade oils. Similar urethane oil films were synthesized by Saravari et al. [52] using palm oil and toluene diisocyanate. They had good characteristic features such as hardness, water and acid resistance, good impact strength and good adhesion. Depolymerized PET waste has also been used to synthesize block copolymers [53,54]. This development was considered commercially important because of two reasons: (a) it gives value addition to the waste; (b) the polyester polyols thus formed had terephthalate-repeating units which eliminated the trouble of sublimation arising from the direct utilization of TPA as starting material in the polyol synthesis [50]. Multiblock copolyesters were also synthesized using glycolyzed PET waste [55].

4.6.2 Unsaturated Polyesters

Unsaturated polyester resins are extensively used as matrix for fiber-reinforced composites as replacement to traditional structural materials such as steel, wood in a range of engineering applications. Fiber-reinforced plastics were made of cross-linkable polymers (usually unsaturated polyesters, epoxy resins) and reinforcing fibers (chiefly glass, carbon, aramid, boron, quartz, high molecular-weight polyethylene, ceramic and hybrid combinations) [56]. These composites are high performing and present excellent chemical resistance, weather-ability, ease of fabrication and high strength to weight ratio. They can either be synthesized from virgin material [57] or from recycled products of PET waste [58]. Resin created from waste PET is useful in making polymer mortars and polymer concretes [59] and in recent times was applied for manufacturing modified unsaturated polyesters with enhanced thermal [60] and mechanical characteristics [61]. Pimpan et al. [62] reported the synthesis and curing performance of unsaturated polyester resins from postconsumer PET bottles. The results showed that the properties of synthesized resin did not depend on the type of bottle thus no separation of bottles is required. PET based epoxy-resin with excellent adhesion properties were synthesized in another study [63]. This resin was used to prepare another set of epoxy ester resin, which when tailored using melamine–formaldehyde and urea–formaldehyde converted to epoxy ester amino resins. These epoxy films have shown considerably better acid and alkaline resistance.

Medium-oil alkyd resins were also obtained by simultaneously utilizing crude glycerol and waste PET as raw materials [64]. PET was depolymerized with EG in the presence of sodium bicarbonate and then two alkyd resins were obtained from monoglycerides, water soluble and water insoluble and phthalic anhydride. The resins obtained exhibit improved drying degree, thermal stability, and hardness.

4.6.3 Acrylate/Methacrylate-Terminated products

To comply with latest solvent emissions guidelines, coatings now a day are cured using either electron beam or ultraviolet radiations [3]. Such processes are green as they do not make use of any volatile organic solvents. Particular applications include, printing ink, metal and wood coatings. Such formulations are made using liquid monomers (acrylates being most common) which polymerize when irradiated. Their composition consists of an acrylate terminated urethane oligomer, a multifunctional acrylate for cross-linking and a mono-acrylate reactive diluent [65].

PET waste when transformed to either acrylate or methacrylate-terminated PET oligomers [66] gives viscous oils that could be mixed with commercial radiation-curable oligomers, cast into films and cured by UV irradiation. These outcomes increase the possibility of converting PET waste into high value-added radiation-curable oligomers meant for coatings applications.

4.7 Conclusion

Glycolysis is the breakdown of ester linkages of PET using a glycol, producing oligomers with hydroxyl as terminal group. These oligoesters find their uses as starting material for a number of applications like in making of polyurethanes, unsaturated polyesters, and saturated polyester plasticizers. The glycolyzed PET products further have the prospect of application in paints and UV curable coatings meant for wood surfaces.

References

[1] Cischem. Com Co., Ltd. Publication, The world market analysis of PET value chain, 2010.

[2] A. Drah, J.D. Rusmirovic, M. Milosevic, M. Kalifa, I. Stojiljkovic, M. Rancic, et al., Techno-economic analysis of unsaturated polyester production from waste PET, Zastita Materijala 57 (4) (2016) 605–612.

[3] M.S. Farahat, D.E. Nikles, New motivation for the depolymerization products derived from poly(ethylene terephthalate) (PET) waste: a review, Macromol. Mater. Eng. 290 (2005) 13–30.

[4] S.M. Al-Salem, Establishing an integrated databank for plastic manufacturers and converters in Kuwait, Waste Manag. 29 (1) (2009) 479–484.

[5] S.M. Al-Salem, P. Lettieri, J. Baeyens, Recycling and recovery routes of plastic solid waste (PSW): a review, Waste Manag. 29 (10) (2009) 2625–2643.

[6] F. Awaja, D. Pavel, Recycling of PET, Eur. Polym. J. 41 (7) (2005) 1453–1477.

[7] M.M. Fisher, Plastics recycling, in: A.L. Andrady (Ed.), Plastics and Environment., John Wiley and Sons, 2003, pp. 563–627.

[8] J. Scheirs, Polymer Recycling: Science, Technology and Applications, John Wiley & Sons Ltd, Journals, Baffins Lane, Chichester, 1998, p. 591.

[9] D. Paszun, T. Spychaj, Chemical recycling of poly(ethylene terephthalate), Ind. Eng. Chem. Res. 36 (4) (1997) 1373–1383.

[10] M.J. Collins, S.H. Zeronian, M.L. Marshall, Analysis of the molecular weight distributions of aminolyzed poly(ethylene terephthalate) by using gel permeation chromatography, J. Macromol. Sci.—Chem. 28 (8) (1991) 775–792.

[11] Blackmon K.P., Fox D.W., Shafer S.J. (1988) Eur Patent 365,842.

[12] V. Sinha, M.R. Patel, J.V. Patel, PET waste management by chemical recycling: a review, J. Polym. Environ. 18 (1) (2010) 8–25.

[13] D. Pant, Polycarbonate waste management using glycerol, Process Safety and Environmental Protection 100 (2016) 281–287.

[14] U.R. Vaidya, V.M. Nadkarni, Unsaturated polyesters from PET waste: kinetics of polycondensation, J. Appl. Polym. Sci. 34 (1) (1987) 235–245.

[15] S.J. Chiu, W.H. Cheng, Thermal degradation and catalytic cracking of poly(ethylene terephthalate), Polym. Degrad. Stab. 63 (3) (1999) 407–412.

[16] G. Xi, M. Lu, C. Sun, Depolymerization of waste PET into monomer of BHET, Polym. Degrad. Stab. 87 (1) (2005) 117–120. ISSN 0141-3910.

[17] K. Troev, G. Grancharov, R. Tsevi, I. Gitsov, A novel catalyst for the glycolysis of poly(ethylene terephthalate), J. Appl. Polym. Sci. 90 (4) (2003) 1148–1152.

[18] S. Shukla, K. Kulkarni, Depolymerization of poly(ethylene terephthalate) waste, J. Appl. Polym. Sci. 85 (8) (2002) 1765–1770.
[19] S. Shukla, V. Palekar, N. Pingale, Zeolite catalyzed glycolysis of poly (ethylene terephthalate) bottle waste, J. Appl. Polym. Sci. 110 (1) (2008) 501–516. ISSN 1097-4628.
[20] M. Imran, et al., Metal–oxide-doped silica nanoparticles for the catalytic glycolysis of polyethylene terephthalate, J. Nanosci. Nanotechnol. 11 (1) (2011) 824–828.
[21] M. Imran, et al., Manganese-, cobalt-, and zinc-based mixed-oxide spinels as novel catalysts for the chemical recycling of poly(ethylene terephthalate) via glycolysis, Polym. Degrad. Stab. 98 (4) (2013) 904–915.
[22] G. Park, et al., One-step sonochemical synthesis of a graphene oxide–manganese oxide nanocomposite for catalytic glycolysis of poly(ethylene terephthalate), Nanoscale 4 (13) (2012) 3879–3885.
[23] Q.F. Yue, et al., Glycolysis of poly(ethylene terephthalate) (PET) using basic ionic liquids as catalysts, Polym. Degrad. Stab. 96 (4) (2011) 399–403.
[24] A.M. Al-Sabagh, et al., Glycolysis of poly(ethylene terephthalate) catalyzed by the Lewis base ionic liquid [Bmim][OAc], Ind. Eng. Chem. Res. 53 (48) (2014) 18443–18451.
[25] M. Ghaemy, K. Mossaddegh, Depolymerisation of poly(ethylene terephthalate) fibre wastes using ethylene glycol, Polym. Degrad. Stab. 90 (3) (2005) 570–576.
[26] J.Y. Chen, et al., Depolymerization of poly(ethylene terephthalate) resin under pressure, J. Appl. Polym. Sci. 42 (6) (1991) 1501–1507.
[27] C. Chen, Study of glycolysis of poly(ethylene terephthalate) recycled from postconsumer soft-drink bottles. III. Further investigation, J. Appl. Polym. Sci. 87 (12) (March 2003) 2004–2010. ISSN 1097-4628.
[28] A.S. Goje, S. Mishra, Chemical kinetics, simulation, and thermodynamics of glycolytic depolymerization of poly(ethylene terephthalate) waste with catalyst optimization for recycling of value added monomeric products, Macromol. Mater. Eng. 288 (4) (2003) 326–336.
[29] J.R. Campanelli, M.R. Kamal, D.G. Cooper, Kinetics of glycolysis of poly(ethylene terephthalate) melts, J. Appl. Polym. Sci. 54 (11) (1994) 1731–1740.
[30] S. Shukla, A. Harad, Glycolysis of polyethylene terephthalate waste fibers, J. Appl. Polym. Sci. 97 (2) (2005) 513–517.
[31] U. Heiz, U. Landman, Nanocatalysis, Springer-Verlag, New York, 2007, 978-1-61583-152-4.
[32] M. Niederberger, N. Pinna, Metaloxide Nanoparticles in Organic Solvents: Synthesis, Formation, Assembly and Application., Springer Science & Business Media, 2009.
[33] L. Bartolome, et al., Recent Developments in the Chemical Recycling of PET, INTECH Open Access Publisher, 2012.

[34] H. Wang, et al., Glycolysis of poly(ethylene terephthalate) catalyzed by ionic liquids, Eur. Polym. J. 45 (5) (2009) 1535–1544.
[35] H. Wang, et al., Degradation of poly(ethylene terephthalate) using ionic liquids, Green Chem. 11 (10) (2009) 1568–1575.
[36] Q. Wang, et al., Urea as an efficient and reusable catalyst for the glycolysis of poly(ethylene terephthalate) wastes and the role of hydrogen bond in this process, Green Chem. 14 (9) (2012) 2559–2566.
[37] M. Imran, et al., Sub- and supercritical glycolysis of polyethylene terephthalate (PET) into the monomer bis(2-hydroxyethyl) terephthalate (BHET), Polym. Degrad. Stab. 95 (9) (2010) 1686–1693.
[38] N.D. Pingale, S.R. Shukla, Microwave assisted ecofriendly recycling of poly (ethylene terephthalate) bottle waste, Eur. Polym. J. 44 (2008) 4151–4156.
[39] J.Y. Cho, C.-J. Hong, H.-M. Choi, Microwave-assisted glycolysis for PET with highly hydrophilic surface, Ind. Eng. Chem. Res. 52 (6) (2013) 2309–2315.
[40] S. Chaudhary, et al., Microwave assisted glycolysis of poly(ethylene terepthalate) for preparation of polyester polyols, J. Appl. Polym. Sci. 129 (5) (2013) 2779–2788.
[41] G. Giannotta, et al., Processing effects on poly(ethylene terephthalate) from bottle scraps, Polym. Eng. Sci. 34 (15) (1994) 1219–1223.
[42] M. Paci, F.P. La Mantia, Influence of small amounts of polyvinylchloride on the recycling of polyethyleneterephthalate, Polym. Degrad. Stab. 63 (1) (1999) 11–14.
[43] F. Villain, J. Coudane, M. Vert, Thermal degradation of polyethylene terephthalate: study of polymer stabilization, Polym. Degrad. Stab. 49 (3) (1995) 393–397.
[44] A.M. Haghighi, M. Khoonkari, Y. Sefidbakht, K. Shekoohi, A. Ghaderian, Chemical recycling of PET wastes with different catalysts, Int. J. Polym. Sci. (2015) 1–11.
[45] R. López-Fonseca, et al., Kinetics of catalytic glycolysis of PET wastes with sodium carbonate, Chem. Eng. J. 168 (1) (2011) 312–320.
[46] F.-A. El-Toufaili, G. Feix, K.-H. Reichert, Mechanistic investigations of antimony-catalyzed polycondensation in the synthesis of poly(ethylene terephthalate), J. Polym. Sci. A: Polym. Chem. 44 (3) (2006) 1049–1059.
[47] Ch-H. Chen, Ch-Y. Chen, Yu-W. Lo, Ch-F. Mao, W.-T. Liao, Studies of glycolysis of poly(ethylene terephthalate) recycled from postconsumer soft-drink bottles. II. Factorial experimental design, J. Appl. Polym. Sci. 80 (7) (2001) 956–962.
[48] M. Kacperski, S. Tadeusz, Rigid polyurethane foams with poly(ethylene terephthalate)/triethanolamine recycling products, Polym. Adv. Technol. 10 (10) (1999) 620–624.
[49] I. Ristić, et al., Investigation of potentional use of recycled poly(ethylene terephthalate) in polyurethane synthesis, Contemp. Mater. 1 (3) (2012) 86–92.

[50] G. Macijauskas, V. Jankauskaite, Epoxy resin and polyurethane composition from glycolized poly(ethylenephthalate) wastes, Mater. Sci. 19 (3) (2013). ISSN 1392-1320.
[51] A. Akar, O.G. Atici, Y. Ayar, O. Mercit, Synthesis of block copolymers via redox polymerization, J. Appl. Polym. Sci. 71 (1999) 1385.
[52] O. Saravari, K. Pathomwattanasak, V. Pimpan, Synthesis of urethane oils from waste poly(ethylene terephthalate) bottles, J. Appl. Polym. Sci. 105 (2007) 648–653.
[53] O.G. Atici, et al., Synthesis of block copolymers via redox polymerization, J. Appl. Polym. Sci. 71 (9) (1999) 1385–1395.
[54] N. Kizilcan, et al., Block copolymers of styrene containing oligomeric esters of terephthalic acids, J. Appl. Polym. Sci. 76 (5) (2000) 648–653.
[55] A. El Mejjatti, et al., Chemical recycling of poly(ethylene terephthalate). Application to the synthesis of multiblockcopolyesters, Express Polym. Lett 8 (2014) 544–553.
[56] D.E. Nikles, M.S. Farahat, New motivation for the depolymerization products derived from poly(ethylene terephthalate) (PET) waste: a review, Macromol. Mater. Eng. 290 (2005) 13–30.
[57] A.A. Abdel-Azim, B.H. Mahmoud, M.S. Farahat, Synthesis of unsaturated polyester resins for large sized, Compos. Polym. Adv. Technol. 5 (5) (1994) 269–274.
[58] M. Lu, S. Kim, Unsaturated polyester resins based on recycled PET: preparation and curing behavior, J. Appl. Polym. Sci. 80 (7) (2001) 1052–1057.
[59] D.J. Suh, O.O. Park, K.H. Yoon, The properties of unsaturated polyester based on the glycolyzed poly(ethylene terephthalate) with various glycol compositions, Polymer 41 (2) (2000) 461–466.
[60] M.S. Farahat, A.-A.A. Abdel-Azim, M.E. Abdel-Raowf, Modified unsaturated polyester resins synthesized from poly(ethylene terephthalate) waste. 1. Synthesis and curing characteristics, Macromol. Mater. Eng. 283 (1) (2000) 1–6.
[61] M.S. Farahat, Mechanical characteristics of modified unsaturated polyester resins derived from poly(ethylene terephthalate) waste, Polym. Int. 51 (2) (2002) 183–189.
[62] V. Pimpan, R. Sirisook, S. Chuayjuljit, Synthesis of unsaturated polyester resin from postconsumer PET bottles: effect of type of glycol on characteristics of unsaturated polyester resin, J. Appl. Polym. Sci. 88 (3) (2003) 788–792.
[63] G. Güçlü, C. Cigdem, A. Bal, Synthesis and film properties of epoxy esters modified with amino resins from glycolysis products of postconsumer PET bottles, Polym. Eng. Sci. 55 (11) (2015) 2519–2525.
[64] N.S. Todorov, M.F. Radenkov, D.D. Todorova, Utilization of crude glycerol and waste polyethylene terephthalate for production of alkyd resins, J. Chem. Technol. Metall. 50 (3) (2015) 240–248.

[65] M.S. Farahat, D.E. Nikles, Value-added radiation-curable oligomers from waste PET, Radtech Rep. 17 (4) (2003) 39–42.
[66] K.-D. Ahn, Ch-M. Chung, Y.-H. Kim, Synthesis and photopolymerization of multifunctional methacrylates derived from bis-GMA for dental applications, J. Appl. Polym. Sci. 71 (12) (1999) 2033–2037.

Further Reading

J.-W. Chen, L.-W. Chen, W.-H. Cheng, Kinetics of glycolysis of polyethylene terephthalate with zinc catalyst, Polym. Int. 48 (9) (1999) 885–888.
Q. Wang, et al., Deep eutectic solvents as highly active catalysts for the fast and mild glycolysis of poly(ethylene terephthalate) (PET), Green Chem. 17 (4) (2015) 2473–2479.

5 Depolymerization of PET Bottle via Methanolysis and Hydrolysis

Myungwan Han
Department of Chemical Engineering & Applied Chemistry, Chungnam National University, Daejeon, Korea

5.1 Introduction

Polyethylene terephthalate (PET) is a saturated polyester composed of its constituent raw materials: terephthalic acid (TPA) and ethylene glycol (EG) or dimethyl terephthalate (DMT) and EG. The PET is extensively used for its outstanding thermal stability, clarity, strength, and moldability in the production of films, bottles, and fibers. Recent growing production and consumption of the PET bottles generated an increased amount of PET waste. Dumping of the waste PET bottles has posed serious environmental as well as economic problems. Clear PET bottle can be easily recycled through a simple process of washing and remelting and used in new PET products. However, contaminated or colored bottle cannot be recycled in the same manner. Low quality PET bottle, colored or contaminated bottle, can be recycled through a chemical break down of the PET into its monomers, purification of the monomers and conversion into new PET.

PET can be depolymerized on various ways to yield monomers such as TPA, DMT, and bis-2-hydroxyethyl terephthalate (BHET) for the production of fresh polyester. Depolymerization agent classifies the depolymerization methods: methanolysis, glycolysis, hydrolysis, etc. EG is used for glycolysis of PET which yields BHET and oligomers. Methanol is used for methanolysis of PET which produces DMT and EG. Methanolysis can take a crucial role of depolymerizing low quality polyester bottles. Hydrolysis can produce TPA and EG, i.e., the monomers from which PET is directly produced in a TPA-based PET production plant. Fig. 5.1 summarizes the main chemical depolymerization reactions used in recycling of PET bottle. The choice of process should be determined considering starting material and the demand of depolymerized

Recycling of Polyethylene Terephthalate Bottles.
DOI: https://doi.org/10.1016/B978-0-12-811361-5.00005-5

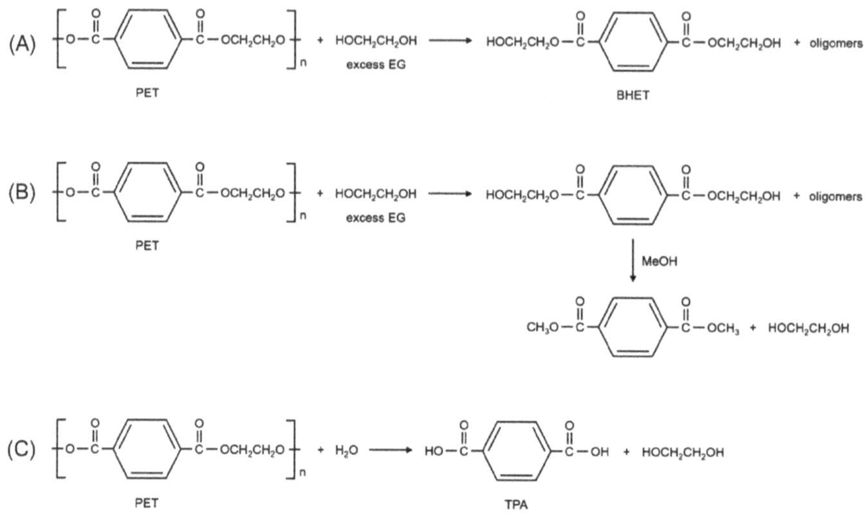

Figure 5.1 Depolymerization of PET: (A) glycolysis; (B) methanolysis; (C) hydrolysis.

products. Depolymerization kinetics of PET are indispensable for building a commercially viable depolymerization process. In this chapter, various PET degradation alternatives based on methanolysis and hydrolysis methods are reviewed.

5.2 Depolymerization of Polyethylene Terephthalate Bottles

Chemical recycling through depolymerization of PET into its monomer is advantageous because permanent recycling of PET could be achieved. Various depolymerization methods of PET, such as methanolysis in liquid or vapor methanol, glycolysis in liquid EG, and hydrolysis under the existence of an alkali, have been developed on the commercial and pilot scale.

5.2.1 Glycolysis

Glycolysis is the most cost-effective and commercially viable process for recycling PET bottles chemically. Glycolysis uses excess glycol, commonly EG, to produce BHET and its oligomers under pressure and temperature in the range of 180–240°C [1–4]. The mechanism of glycolysis consists of the following steps. Glycol diffuses into the polymer and swells the polymer. The diffused glycol reacts with an ester bond in the chain. The swelling increases the diffusion rate as well as the reaction rate of the glycolysis. The reaction rate increases with the polymer surface area, it is better to make the size of the PET chip smaller by grinding, cutting, etc. BHET is generally purified by melt filtration under pressure and treatment with activated carbon to remove impurities and color. Usually, the glycolysis method is used for recycling high quality PET bottles.

Partial glycolysis can be employed as a pretreatment step where PET molecules are attacked by EG molecules and reduced to oligomers with low molecular weight. The glycolysis reaction of PET bottle and EG is mostly achieved by extrusion. The low weight oligomer can be used as a feedstock to methanolysis or hydrolysis steps, or repolymerization process.

5.2.2 Methanolysis

PET methanolysis produces DMT and EG, which are also raw materials required for the production of PET. Methanolysis is more expensive than glycolysis. However, methanolysis can treat low quality feedstocks, reducing the feedstock costs since methanolysis is more tolerant of contamination [5].

5.2.2.1 Liquid Methanolysis

PET methanolysis depolymerizes PET bottles with methanol at temperature 180–280°C and pressure 20–40 atm [6–9]. Typical transesterification catalysts, such as zinc acetate, magnesium acetate, cobalt acetate, and lead dioxide, are usually used for the reaction. The methanolysis occurs at high pressure so that the methanol can be maintained as a liquid during the reaction. PET can be directly subjected to methanol or glycolysis and subsequent methanolysis. The yield of DMT is in the range of 80% and 85%. The products of the methanolysis are usually separated and purified by distillation or crystallization. Purified DMT can be reused for PET polymerization. The methanolysis process can be operated under both batch and continuous conditions. The continuous process requires a complex apparatus to introduce PET wastes into the methanolysis reactor working under high pressure.

5.2.2.2 Vapor Methanolysis

Superheated vapor methanol can be used instead of liquid methanol for transesterification of the oligomer or crude BHET, which are obtained by glycolysis, by bubbling the methanol vapor through the glycolysis product. Naujokas and Ryan [10] proposed a methanolysis process in which oligomers of the DMT and EG are used for dissolving scrap PET waste and superheated methanol is sparged into the resulting solution. The reaction mixture is held at lower pressures than necessary to keep the methanol a liquid, which makes it easier to remove the resultant DMT as a vapor. The process can treat more contaminated PET than conventional liquid-phase methanolysis by removing esters and alcohols as vapors [5]. The reaction equilibrium is also shifted by vapor removal of the products from the reactor, promoting the reaction conversion and leading to almost complete conversion. Therefore, vapor methanolysis shows a higher reaction yield than liquid-phase methanolysis, although reaction rate is much faster in liquid methanolysis than in vapor methanolysis.

The reaction usually occurs in an agitated tank where the methanol gas is passed through the liquid solution. For the two-phase reaction, intimate mixing between the phases should be carried out to facilitate mass transfer of the reactants between phases. There may be practical reasons to use a sparged agitated vessel. The agitator helps the gas to be dispersed as small bubbles and the agitation of the liquid allows good contact between the gas and the liquid [11]. Gamble et al. [12] proposed a staged column reactor for the depolymerization and for separating monomer components from the higher boiling materials. The influence of temperature on gas–liquid reactions is more complex than homogenous reactions, in this case liquid methanolysis. As the temperature increases, rate of reaction increases but solubility of the methanol gas in the liquid decreases, contributing to decrease the reaction rate. The influence of pressure is also rather complex, because increasing pressure accelerates the reaction rate but decreases removal rate of the products, DMT and EG. Some of these effects contribute to increasing the reaction rate. Others have a decreasing effect. Kim and Han [13] examined the effects of parameters such as methanol feeding rate, methanol purity, reaction pressure, reaction temperature, and amount of catalyst on DMT production yield in vapor methanolysis reactor. DMT production yield goes up with methanol feeding rate and gives a maximum at reaction temperature 250–260°C and reaction pressure 3–6 atm. Vapor methanolysis is currently successfully applied to PET scrap.

Vapor removal of volatile PET monomers can also be used to provide monomer to form BHET and related species in transesterification reactor without much separation efforts. Separation costs are high so that skipping the separation units makes overall operation of the plant economic. The product, BHET and related monomers, can be used as raw materials in either TPA based or DMT based production plants. The Eastman Chemical Company has patented a hybrid methanolysis/BHET depolymerization process [14–16].

5.2.2.3 Supercritical Methanolysis

Recently, supercritical methanol at a temperature of 300°C and pressure above 80 atm has been used to carry out PET methanolysis. Reactivity of supercritical methanol has received an attention for depolymerization of PET. The supercritical fluid has high density over its critical point, such as in liquid state, and high kinetic energy as in a gas molecule. PET decomposition was much faster under these conditions than when using liquid methanol. Sako et al. [17] reported that supercritical methanolysis depolymerizes PET into monomers in 30 min without catalyst. Goto et al. [18,19] investigated the reaction mechanism and kinetic of PET depolymerization. Genta et al. [20] studied PET depolymerization using supercritical methanol as depolymerization agent. The rate of PET depolymerization in supercritical methanol was found to be greater than that of PET depolymerization in vapor methanol. Yang et al. [21] reported that temperature, weight ratio of methanol to PET, and reaction time were essential parameters for both yield of DMT and degree of depolymerization, but the pressure was not considerable above critical point of methanol. Liu et al. [22] investigated PET methanolysis under supercritical conditions using response surface methodology and reported that ratio of methanol to PET mass is not significant parameter in comparison to reaction temperature and time. In supercritical methanolysis, the depolymerization rate is much faster than in vapor or liquid methanolysis. However, the severe reaction conditions make the process more cost intensive and also cause continuous operation of the process very difficult.

5.2.2.4 Hydrolysis of Dimethyl Terephthalate

TPA has been used instead of DMT as the raw material in recently built PET production processes. DMT production is not useful for the TPA-based PET production process. Hydrolyzing the DMT with water into TPA is necessary to utilize methanolysis in TPA-based PET production. The hydrolysis is generally carried out above melting point of DMT, about 150°C, but below the temperature at which significant DMT degradation occurs, usually about 380°C and at a pressure sufficient to maintain a liquid phase in the bottom of the hydrolysis reactor, about 17–100 bar [23]. The product TPA is crystallized in a single or multistage crystallization process. Integrating the hydrolysis of DMT to TPA to the methanolysis process needs an additional considerable cost. Sim and Han [24] studied the kinetics of the hydrolysis reaction. The reaction was found to be a reversible reaction limited by reaction equilibrium. Kim et al. [25] proposed a new hydrolysis reactor using reactive distillation column based on Sim and Han's work. The process continuously removes methanol generated from the hydrolysis reaction zone at the top of the column, shifting the reaction equilibrium and makes it possible to get almost complete conversion of DMT to TPA.

5.2.2.5 Purification of Dimethyl Terephthalate and Ethylene Glycol

Purification cost to get pure DMT and EG is greater than the cost for the methanolysis reaction. Methanolysis of postconsumer PET produces an extensive mixture of glycols, alcohols and phthalate derivatives. The separation and refinement of these make methanolysis a rather costly process [5]. It is necessary that catalysts are deactivated or kept out of the reactants before the reaction mixture is introduced into subsequent separation stages of the process. Distillation can be used to separate EG from the DMT. Crystallization and/or distillation can separate and purify DMT. DMT and EG form an azeotrope, which should be considered when distillation is used for separating DMT–EG mixture. Azeotropic distillation of DMT and EG has been patented [26,27]. DMI from an isophthalic component in PET bottle polymer has similar boiling point as DMT and is not easily separated by distillation. Operation results of commercial process, i.e., Teijin process, indicated that crude DMT obtained from the methanolysis can be purified to a quality comparable to that of virgin DMT. The purified DMT has a purity that can be used for the production of PET in the transesterification process.

5.2.3 Hydrolysis

Water can be used to break polyester chains into TPA and EG. The depolymerized product can be used directly in TPA-based PET synthesis plant. The process can be carried out under neutral, acidic or basic conditions. Hydrolysis of PET chain produces hydroxyl and carboxylic end group. One molecule of water attacks the chain and causes a scission of the chain, creating a carboxylic end group.

$$\sim C(O) - C_6H_4 - C(O) - O - CH_2CH_2 - O\sim + H_2O \rightarrow \sim C(O) - C_6H_4 - C(O) - OH + HO - CH_2CH_2 - O\sim$$

However, hydrolysis is not still commercially viable option although it has been actively studied in recent years. One major obstacle to commercialization is a purification problem of recycled TPA. Purifying TPA from the reaction mixture is difficult due to its low solubility and low vapor pressure. Some impurities in PET wastes, which are not soluble in hydrous agents, are not separated by trans-crystallization using acetic acid/water.

5.2.3.1 *Acid Hydrolysis*

Several patents describe the acidic hydrolysis of PET with concentrated sulfuric acid (>14.5 M) at temperatures of 25–100°C and atmospheric pressure [28–30]. The hydrolysis product is neutralized with sodium hydroxide, which causes the formation of the corresponding TPA sodium salt, which is soluble in water. In the final stage of the process, the resulting solution is again acidified to reprecipitate TPA, which is obtained with a purity >99%. In this process, it is necessary to separate large amounts of concentrated sulfuric acid and purify EG containing sulfuric acid, which makes the process very expensive. The reaction mixture is corrosive and large amount of liquid wastes, containing inorganic salts and sulfuric acid, are generated. Yoshioka et al. [31] investigated the effect of sulfuric acid concentration on the acid hydrolysis of PET and developed a process using sulfuric acid of low concentration. The degradation of PET shows a sharp increase above 5 M and almost complete around 7 M. After the reaction, separation of TPA from the unreacted PET can be achieved by dissolving with ammonia (5 M). Yoshioka et al. [32] describe a hydrolysis of PET bottle using nitric acid (7–13 M) at temperatures of 70–100°C for 72 h. TPA and EG are obtained as reaction products, and EG is partially oxidized to oxalic acid which is more valuable than EG.

5.2.3.2 *Alkaline Hydrolysis*

The alkaline hydrolysis of PET involves treating the polyester with an aqueous solution of sodium hydroxide (4–20 wt%) under pressure at temperatures between 200 and 250°C for periods of several hours [33,34]. Distillation is used to separate the reaction mixture, which includes ethan diol (ED) and a heat-stable salt (disodium terephthalate), and recover the ED. The heat stability of the terephthalate salt makes the process different from other chemical recycling methods because the salt is easily separated and purified at ambient pressure and low temperature [5]. Alkaline hydrolysis can deal with highly contaminated, postconsumer PET with relatively simple process. The process requires less capital and operation costs than methanolysis since the process operates at lower temperatures and pressures than methanolysis [35]. Neutralizing the reaction mixture with a strong acid yields pure TPA.

Yoshioka et al. [31] describe an alkali hydrolysis process in which TPA and oxalic acid can be obtained from PET wastes in a concentrated NaOH solution. Heated aqueous NaOH solution converts PET into sodium terephthalate and EG. Oxygen is introduced into the solution, converting EG to oxalic acid and CO_2. TPA yield is almost 100% because the sodium terephthalate is not affected by the oxidation. The oxidation reaction contributes to improving the process economics since oxalic acid is more valuable product than EG. It is notable that the oxidation step can remove the green color in green PET bottle. TPA impurities in PET alkaline hydrolysis can be significantly reduced by introducing the oxidation step of impurities and converting them into insoluble forms.

In recent years, several works have appeared on the alkali degradation of PET in solvents other than water. Collins and Zeronian [36] used methanolic sodium hydroxide as solvent and obtained a faster degradation than when using aqueous sodium hydroxide. Hu et al. [37] employed sodium hydroxide dissolved in anhydrous EG to degrade PET and obtained disodium terephthalate as a major product. TPA was easily recovered by dissolution in water of the sodium salt followed by acidification with HCl. The authors showed that potassium hydroxide dissolved in ethanol and methanol can be used for alkaline hydrolysis.

Different ethers, such as dioxane, tetrahydrofuran, and 1,2-dimethyloxyethane, as cosolvents in nonaqueous alkali solutions can be used to accelerate the degradation of PET. It can be explained that the ether part will take a role of swelling of the PET solid and the alcoholic part will help NaOH or KOH attack the chemical structure of PET [38].

5.2.3.3 Neutral Hydrolysis

Hydrolytic scission of PET with water or steam can be carried out to yield TPA and EG at neutral pH and temperatures above the melting points under high pressure condition [39]. PET hydrolysis rate is much greater in molten state than as a solid. Campanelli et al. [4] reported that complete depolymerization could be achieved within 2 h at 265°C with an initial reactant ratio (water/PET) of 5.1. The hydrolysis of simple ester in aqueous solution is catalyzed by metal salts. Campanelli et al. [40] reported that zinc acetate and sodium acetate increase the rate of PET hydrolysis by around 20% and the catalytic effect of zinc salts as well as sodium salts is caused by electronic destabilization of polymer–water interface in the hydrolysis process. Neutral hydrolysis does not produce unwanted liquid effluents containing inorganic salts and the reaction system for neutral hydrolysis is not corrosive. However, most of the impurities present in the PET are not easily separated from the product TPA, which makes much more intensive purifications necessary for obtaining pure TPA than acid or alkaline hydrolysis [35]. Filtration of the TPA solution dissolved in caprolactam or in aqueous solution of sodium hydroxide [41] or by crystallization of TPA from caprolactam [42] can be used for the purification. EG generated during the reaction can be purified using extraction or distillation.

5.3 Depolymerization Kinetics of Polyethylene Terephthalate

Depolymerization of PET can be thought as a transition process from long polyester chain in solid state to solid oligomers, from solid oligomers to liquid oligomers, and then from liquid oligomers to monomers. Polymer decomposition is a dynamic system where the molecular-weight distribution varies with time owing to chain cleavage. In the degradation of polymers, two reaction paths, random degradation and specific reaction, may contribute to the reaction yielding monomeric species from the polymer [18]. In random scission, binary scission of bonds occurs at any position along the chain. In the specific reaction, scission takes place at the chain end, releasing monomeric species of the polymer. Depolymerization of PET can be carried out with different methods: glycolysis, methanolysis, and hydrolysis, etc. This means that different depolymerization methods can be combined to get a desired monomer from PET bottle.

PET evolves from a solid–liquid heterogeneous reaction to a liquid homogeneous reaction during the glycolysis reaction. Xi et al. [43] suggested that glycolysis of PET experiences at least three steps (oligomer, dimer, and monomer). The glycol diffuses into the polymer, swelling the polymer and the swelling increases the diffusion rate. The diffused glycol attacks ester bonds in the chain, breaking down the chain into smaller fractions. Pardal and Tersac [44,45] investigated evolution of the solid and liquid phases during glycolysis by diethylene glycol (DEG). The evolution was described to be the combined result of diffusion of DEG from liquid phase into the solid polyester (swelling) and diffusion of destructed polyester from solid phase into the liquid phase (dissolution). Viana et al. [46] and Sangalang et al. [47] developed kinetic models for heterogeneous PET glycolysis. The influential reaction parameters on the glycolysis are catalyst concentration, reaction temperature, reaction time [48].

Several researchers studied the kinetics of PET depolymerization in supercritical methanol. Kurokawa et al. [49] proposed a three-step mechanism for PET methanolysis: (1) cleavage at the tie molecule connecting PET crystals; (2) random scission of the polymer into oligomers; and (3) depolymerization of the oligomers to monomers. Genta et al. [20] suggested the kinetic model of PET depolymerization in supercritical methanol that random scission occurs mainly in the heterogeneous phase during the initial stage and specific scission occurs mainly in the homogeneous phase during the final stage. Hwang and Han [50] proposed a new kinetic model to describe the depolymerization of PET chip. In the model, the depolymerization by methanolysis consists of three steps as shown in Fig. 5.2. In the first step, methanol penetrates into the PET particle and random scission occurs making the PET particle to dissolvable solid oligomers. In the second and third step, the solid oligomer is dissolved to liquid oligomer and the dissolved oligomer is converted to DMT and EG. They found that the first step requires longer time than the second and third step. This indicates that it is crucial to shorten the first step to improve the depolymerization efficiency. Genta et al. [20] also claimed that the depolymerization of PET into PET oligomer is more strongly influenced by mass transfer than the depolymerization of PET oligomer into monomers. It follows that partial glycolysis can be used to shorten the first step as a pretreatment step of polyester bottle prior to methanolysis. PET is usually dissolved and undergoes partial glycolysis followed by methanolysis to give DMT and EG.

Genta et al. [20] reported that the depolymerization of PET with a low polymerization degree into its monomers would be a rate determining compared to that with a high polymerization degree and that the depolymerization of PET would proceed consecutively through MHET.

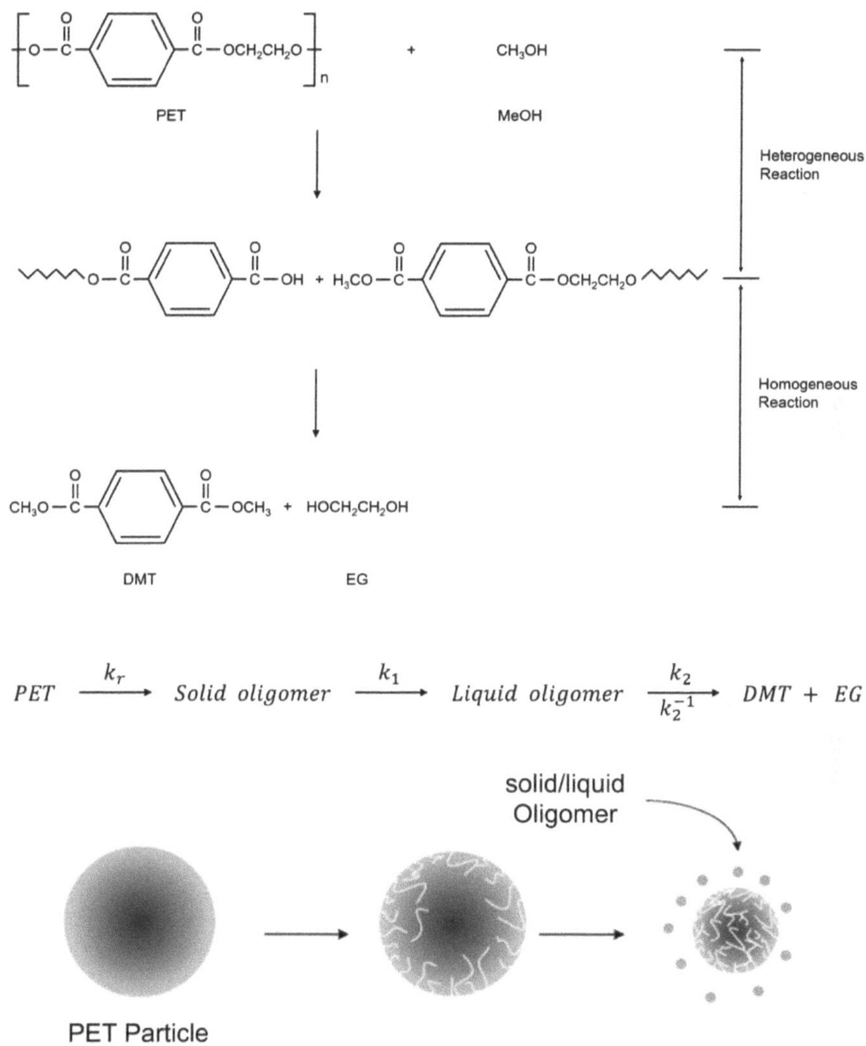

Figure 5.2 Kinetic model for PET methanolysis.

In a batch methanolysis, there exists equilibrium between oligomers and monomer at the end of reaction. This is the main reason why depolymerization reaction slows down at a low polymerization degree. Vapor methanolysis shifts the reaction equilibrium to the direction of forward reaction by removing the products with vapor methanol, which can improve the depolymerization efficiency.

The hydrolysis of PET is thought to be a heterogeneous reaction at temperatures below its melting point and a homogeneous reaction at higher temperatures than melting point. The melting point of PET is around 245–265°C. Since the reaction at temperatures below the melting point proceeds at the solid–liquid interface. Hydrolysis of PET was represented by a modified shrinking core model for the rate determining reaction, in which the effective surface area is assumed to be proportional to the degree of unreacted PET and is affected by the deposition of TPA product on the PET [51,52]. The hydrolysis was facilitated above 5–7 M due to an increased specific surface area of PET by the formation and growth of cracks. Hydrolysis above the melting point makes the reaction occur in homogeneous phase so that complete conversion can be obtained in 30 min, with yields for TPA of 97% and EG of 91%.

5.4 Pros and Cons of Depolymerization Methods

Glycolysis of PET yields BHET as well as its respective oligomers. Glycolysis of PET bottle can be easily added to an existing PET production plant. However, BHET is a waxy solid with a high melting point so that purification by distillation is difficult. The reaction mixture obtained by PET glycolysis is difficult to purify, which makes glycolysis a recycling option limited to PET source of known history and high quality. Main advantage of methanolysis is that DMT is easier to purify than BHET so that the methanolysis process can treat a lower quality PET feedstock than glycolysis. However, methanolysis is more capital and energy intensive than glycolysis. Methanolysis of PET bottle can be thought to consist of heterogeneous reaction and homogeneous reaction. Kinetics study of methanolysis of PET bottle in recent years indicates that limiting step of the depolymerization is a heterogeneous reaction rather than homogeneous reaction. Glycolysis performs better than methanolysis during the heterogeneous depolymerization. Partial glycolysis/methanolysis can be a viable option to treat low quality PET bottles. The majority of all new production of PET is based on purified TPA. TPA can be directly obtained by hydrolysis of PET bottle. However, major disadvantage of producing pure TPA directly from PET bottle is difficulty of removing impurities which are insoluble in hydrous agents. All of the terephthalic acid purification processes are capital-intensive due to the intractable nature of TPA [53]. A promising alternative is depolymerizing PET bottle by methanolysis to get crude DMT, purification of DMT, and subsequent hydrolysis of DMT to PTA. The process is only commercialized recycling path [54].

Table 5.1 Price Premium Required for 25% Recycled Content of PET [53]

Process	Price Premium ($/kg)
Liquid methanolysis	0.112
Vapor methanolysis	0.097
Alkaline hydrolysis	0.149
Neutral hydrolysis	0.121
Methanolysis/BHET	0.082
Glycolysis with color filtering	0.097

Cornell [53] made economic comparison of several processes for recycling mixed colored PET bottles to raw materials which can be used to make new food and beverage bottles. Methanolysis, hydrolysis, methanolysis/BHET hybrid and glycolysis with color filtration can well treat the mixed colored bottles. Table 5.1 shows the price premium required to offset the excessive costs to depolymerize PET bottles in order to sell resin made from the depolymerization processes.

Vapor methanolysis and glycolysis with filtering show better economic potential than alkaline and neutral hydrolysis. This is because separation/purification trains used in hydrolysis and neutral hydrolysis equipment are expensive. The glycolysis with filtration shows low capital investment but operation cost is high, primarily due to product loss in filtration and disposal cost of filtration media. Methanolysis/BHET hybrid appears most promising in economic point of view, which is mainly caused by avoiding cost intensive separation trains.

5.5 History of Feedstock Recycling Techniques From a Standpoint of Patents

Fig. 5.3 shows the development of the technique for feedstock recycling in terms of key patents. Glycolysis was initially patented for BHET recovery. However, glycolysis of PET produces a large amount of oligomers as well as BHET and it is not easy to purify the product BHET. The number of patents regarding glycolysis is not so many. Glycolysis is now widely used as a pretreatment step for hybrid depolymerization process or used for the production of polyols. The key patent for methanolysis began to appear in 1990. Previously, no patent

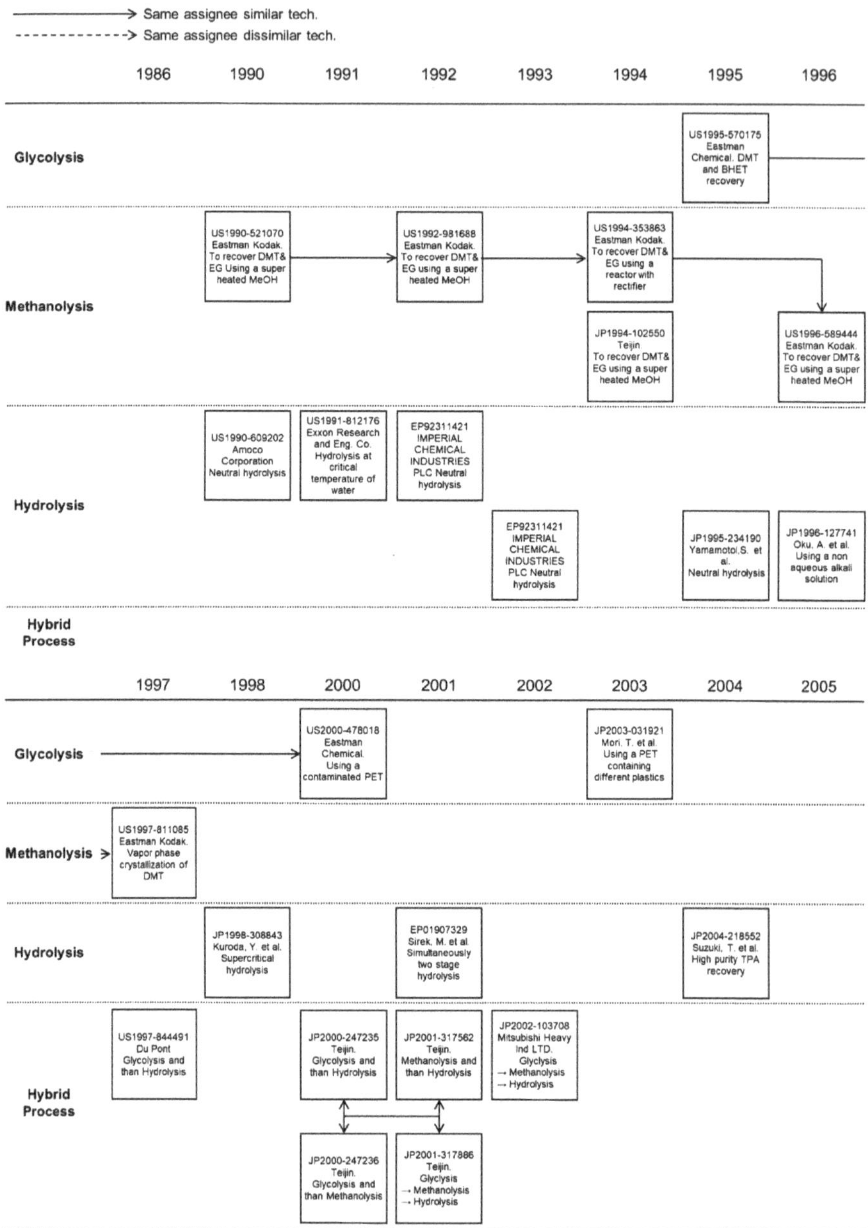

Figure 5.3 Technical flowsheet of polyester depolymerization viewed by the core patents.

application was made because the methanolysis process requires a large investment compared to the glycolysis process in separating and purifying mixtures of glycols, alcohols, phthalate derivatives, etc. From 1990 to 1997, Eastman Kodak filed an active patent application for methanolysis. This is related to the active research of PET recycling at this time. However, since most of the PET manufacturing process uses TPA as raw material monomer rather than DMT, there is no core patent for recovery of DMT using methanolysis process since 1998. Hydrolysis has been patented steadily since 1986. Hydrolysis can be carried out under neutral, acidic or basic conditions. Most of the patents filed were related to neutral hydrolysis. This is because although the neutral hydrolysis requires high temperature and high pressure, there is no corrosion problem which occurs in the acid or base process, and the process is simpler than other hydrolysis processes. Eastman Kodak Co. patented a hybrid depolymerization process including glycolysis followed by methanolysis. This hybrid depolymerization process is intended to take advantage of each of the depolymerization processes described above. Since 1997, patents for hybrid depolymerization process have been actively filed. Recently, a lot of patents have been filed for processes that hybridize hydrolysis with glycolysis and methanolysis process.

5.6 Representative Chemical Recycling Processes Based on Methanolysis and Hydrolysis

5.6.1 Eastman Kodak Process

The process patented by Eastman Kodak Co. [10,12] provides a depolymerization installation which consists of a dissolver, a depolymerization reactor, and a rectifying column. The molten polymer from the reactor and liquid from the rectifying column are used to reduce chain length of the polyester subjected to the dissolver. The polyester then experiences methanolysis by superheated methanol in the reactor. The superheated methanol vapor passes through the reaction mixture in the reactor, heating the reaction mixture to form a melt comprising low molecular weight polyesters and monomers, monohydric alcohol-ended oligomers, glycols, and DMT. The methanol not only acts as a depolymerization agent but also as a carrier gas stream by removing vapor products from reaction mixture. The rectification process separates the vapor products into a gas phase containing the

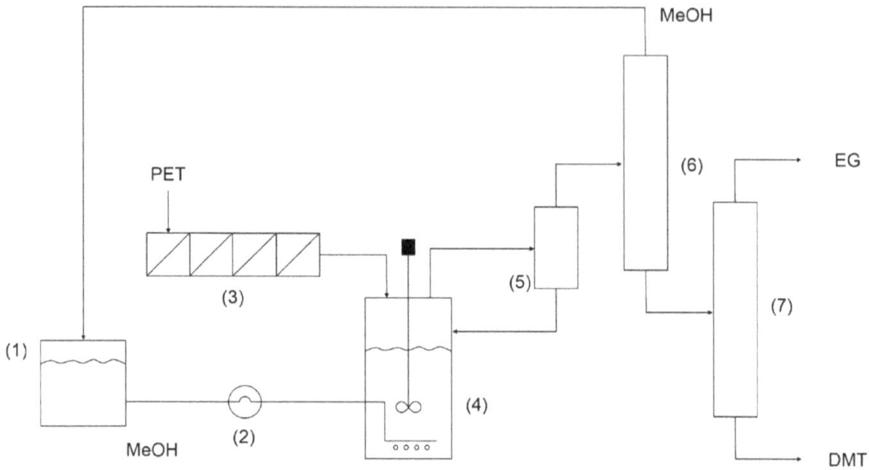

Figure 5.4 Eastman Kodak process: (1) MeOH tank; (2) heater; (3) screw feeder; (4) methanolysis reactor; (5) rectifier; (6) MeOH recovery column; (7) DMT, EG separation column.

monomer components and a liquid phase with oligomers. The gas phase stream is transferred from the rectifier to subsequent separation processes. Fig. 5.4 shows process flowchart for Eastman Kodak process.

5.6.2 Teijin Process

Teijin Ltd. uses its own proprietary decomposition method to break waste PET resin down into raw materials to be used to make textiles and films. Teijin process consists of partial glycolysis and liquid methanolysis followed by hydrolysis. Partial glycolysis produces BHET and its oligomers by depolymerizing recovered PET flakes using EG containing a PET depolymerization catalyst. The depolymerization occurs at a temperature of 175–190°C under a pressure of 0.1–0.5 MPa. The BHET is concentrated by distillation and converted to crude DMT and EG through an ester interchange reaction, and then recrystallized in a MeOH solvent. DMT cakes obtained from recrystallization are further purified by distillation and the purified DMT is hydrolyzed with water to produce TPA at a temperature of 230–250°C. TPA cakes are obtained from the reaction mixture through solid–liquid separation process. Teijin Fiber Ltd. built a plant in 2003 which can produce 50,000 t/y DMT. The products are used for the production of PET fibers and films. Flowchart of Teijin process is shown in Fig. 5.5.

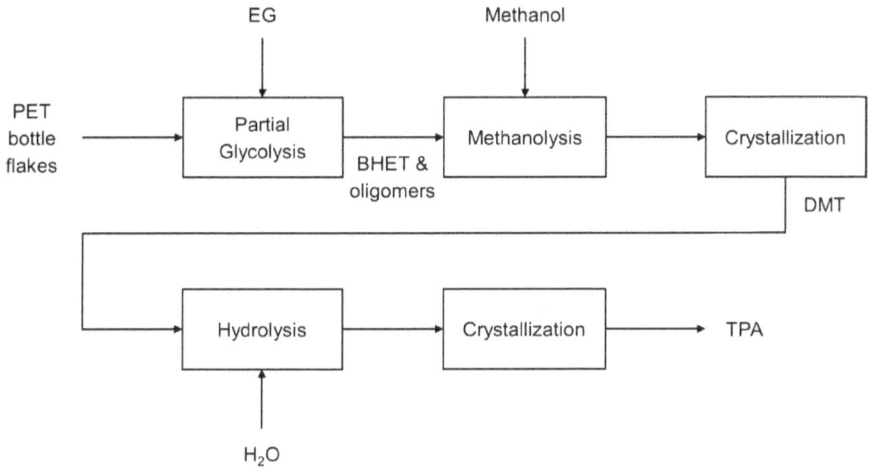

Figure 5.5 Flowchart of Teijin process.

5.6.3 Mitsubishi Process

Mitsubishi Heavy Industries, Ltd., Japan (MHI) has developed a chemical recycling process using supercritical methanol for depolymerizing postconsumer PET bottles into monomers for use as feedstocks for production of PET resin [55,56]. The pilot plant process consists of the following steps: shredding of PET bottle, the depolymerization of the shredded bottle with supercritical methanol, separation and purification of DMT and EG produced from the depolymerization, and the hydrolysis of DMT into TPA. In this process, postconsumer PET bottles are converted into monomers as pure TPA and EG. MHI obtained high purity monomers whose qualities are equivalent to those of virgin monomers using this pilot plant. Postconsumer PET bottles are cut and reduced to flakes in the shredding step. The flakes are depolymerized with supercritical methanol to produce a mixture of DMT and EG, and methanol in the methanolysis step. The mixture is separated and purified mainly by distillation. EG is purified from the mixture of EG and excess methanol in the EG/methanol purification step. The purified DMT monomer is further converted into PTA (highly purified terephthalic acid) in the hydrolysis step. The converted PTA and purified EG are then polymerized in existing PET resin production plants. These steps constitute an ideal cyclical type recycling system. Flowchart of Mitsubishi process is shown in Fig. 5.6.

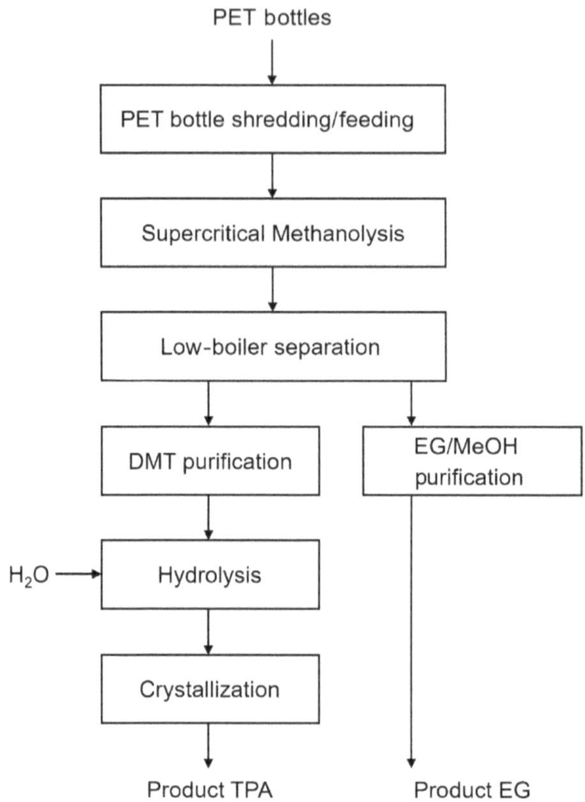

Figure 5.6 Flowchart of Mitsubishi process.

5.6.4 Chungnam National University Process

A new recycling process for getting DMT and EG from PET bottle was developed by Chungnam National University (CNU) [57]. The process is shown in Fig. 5.7. The combined process of partial glycolysis and methanolysis is used to produce DMT and EG. The two processes made a synergy effect to get the depolymerization faster [58]. The proposed CNU methanolysis process consists of two step methanolysis: high pressure semibatch glycolysis/methanolysis and low pressure continuous methanolysis. In high pressure glycolysis/methanolysis, length of polyester chain was reduced by random scission producing oligomers and DMT. The methanolysis reaction rate is high because of highly dissolved concentration of methanol due to high reaction pressure. However, the reaction is limited by equilibrium effect. The equilibrium can be shifted by continuously removing vapor mixture of DMT, EG,

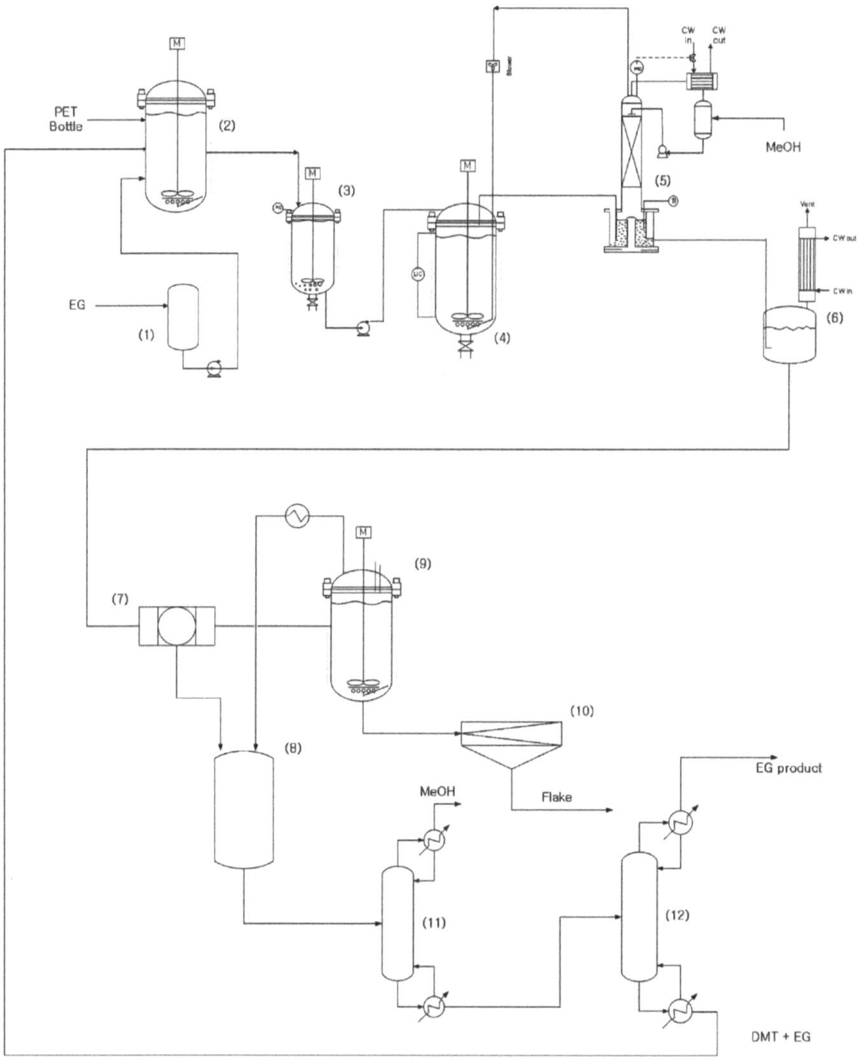

Figure 5.7 Chungnam National University process: (1) EG tank; (2) dissolver; (3) glycolysis/high pressure methanolysis reactor; (4) low pressure methanolysis reactor; (5) rectifier; (6) crystallizer; (7) filter; (8) MeOH storage tank; (9) melter; (10) flake maker; (11) MeOH recovery column; (12) EG recovery column.

and methanol. In low pressure methanolysis reactor, superheated methanol vapor is bubbled into bottom of the reactor in which methanolysis reaction takes place, its products are removed as a vapor mixture. The low pressure methanolysis reactor is integrated with the rectifier where methanol is recovered at the top of the rectifier and recycled to the low pressure methanolysis reactor and bottom products consisting of DMT,

EG and methanol is purified by crystallization or distillation to produce high-purity DMT and EG.

Vapor methanolysis can recycle low-grade polyester bottles to high-purity DMT. Hybrid process combining glycolysis and methanolysis makes depolymerization rate faster than conventional processes. Vapor recycling from rectifier to methanolysis reactor and fresh methanol feeding to rectifier leads to significant energy savings and reduction of conversion cost. Combining methanolysis reactor and rectifier process allows the operating pressure and composition of bottom product in the rectifier easily adjusted so that subsequent separation load can be greatly reduced.

5.7 Conclusions

PET bottle is being more widely used and amount of postconsumer PET bottles is also increasing accordingly. Recycle of the PET bottle to new valuable chemicals becomes more important economically as well as environmentally. In this chapter, we considered various depolymerization methods for chemical recycling of polyester bottle—glycolysis, methanolysis, and hydrolysis. Among the depolymerization methods, methanolysis is most tolerant of low quality PET bottles. However, methanolysis is most capital and energy intensive. We discussed a variety of methanolysis methods such as liquid, vapor, and supercritical methanolysis. Hybridizing other depolymerization methods to methanolysis is crucial to accelerate depolymerization of the polymer and to help reduce energy consumption. For example, partial glycolysis/methanolysis is more efficient than methanolysis only. Partial glycolysis takes a role of decomposing polyester chain into short polymer chains during initial heterogeneous phase and methanolysis depolymerizes the short polymer chain during homogeneous phase. Most recent polyester plants are based on TPA for its feedstock so that DMT product obtained through methanolysis should be converted to TPA by hydrolysis reaction. Hydrolysis can also provide a direct route to TPA from PET bottle, not from DMT. We reviewed various hydrolysis methods: alkaline, acid, and neutral hydrolysis. However, major disadvantage of producing pure TPA directly from PET bottle is difficulty of removing impurities which are insoluble in hydrous agents. Separation and purification of depolymerized products take a great effect on economics of the technology.

References

[1] A. Fujita, M. Sato, M. Murakami, 1986. Process for the depolymerization of polyester scrap. U.S. Patent 4 609 680.

[2] P.L. Johnson, D.A. Teeters, kinetic study of the depolymerization of poly (ethylene terephthalate) recycled from soft-drink bottles, Polym. Prepr. (Am. Chem. Soc., Div. Polym. Chem.) 32 (1991) 144.

[3] J.Y. Chen, Y.C. Ou, C.C. Lin, Depolymerization of poly(ethylene terephthalate) resin under pressure, J. Appl. Polym. Sci. 42 (1991) 1501.

[4] J.R. Campanelli, M.R. Kamal, D.G. Cooper, Kinetics of glycolysis of poly(ethylene terephthalate) melts, J. Appl. Polym. Sci. 54 (1994) 1731.

[5] J. Scheirs, Recycling of PET: Polymer Recycling: Science, Technology And Applications., Wiley Series in Polymer Science, Wiley, Chichester, U.K., 1998.

[6] R. Lotz, G. Wick, C. Neuhaus, 1967. Process for the recovery of dimethyl terephthalate from polyethylene terephthalate. U.S. Patent 3 321 510.

[7] M.N. Marathe, D.A. Dabholkar, M.K. Jain, 1980. Process for the recovery of dimethyl terephthalate from polyethylene terephthalate. GB Patent 2 041 916.

[8] R.E. Michel, 1992. Recovery of methyl esters of aromatic acids and glycols from thermoplastic polyester scrap using methanol vapor. Eur. Patent 484 963.

[9] C. Socrate, R. Vosa, 1995. Continuous process for the recovery of terephthalic acid from waste or used polyalkylene terephthalate polymers. Eur. Patent 662 466; CA 123: 257959u.

[10] A.A. Naujokas, K.M. Ryan, 1991. Recovery process for ethylene glycol and dimethylterephthate. U.S. Patent 5 051 528.

[11] R. Smith, Chemical Process Design andIntegration, John Wiley & Sons, Ltd., 2005.

[12] W.J. Gamble, A.A. Naujoksa, B.R. Debrum, 1994. Process of recovering components from scrap polyester. U.S. Patent 5, 298, 530.

[13] S. Kim, M. Han, A methanolysis process for recycling polyester oligomer waste, Korean Chem. Eng. Res. 42 (1) (2004) 53–58.

[14] M.P. Ekart, T.M. Pell, D.D. Cornell, D.B. Shackelford, 2001. Depolymerization apparatus for recycling polyesters. US Patent 6 191 177.

[15] M.P. Ekart, T.M. Pell, D.D. Cornell, D.B. Shackelford, 1999. Depolymerization process for recycling polyesters. World Patent (WO) 9 920 684 A1.

[16] M.P. Ekart, T.M. Pell, D.D. Cornell, D.B. Shackelford, 2000. Depolymerization process for recycling polyesters. US Patent 6 136 869.

[17] T. Sako, T. Sugeta, K. Otake, Y. Takebayashi, C. Kamizawa, M. Tsugumi, M. Hongo, Kinetic study on depolymerization of poly(ethylene terephthalate) with methanol at high temperature and pressure, Kobunshi Ronbunshu 55 (1998) 685–690.

[18] M. Goto, H. Koyamoto, A. Kodama, T. Hirose, S. Nagaoka, B.J. McCoy, Degradation kinetics of polyethylene terephthalate in supercritical methanol, AIChE J. 48 (2) (2002) 136–141.
[19] M. Goto, M. Genta, 2003. Supercritical methanol for chemical recycling of PET bottle. Proceedings of Second International Symposium on Supercritical Fluid Technology for Energy and Environment Application, 64–73.
[20] M. Genta, T. Iwaya, M. Sasaki, M. Goto, T. Hirose, The stability of PET depolymerization products in supercritical methanol, Ind. Eng. Chem. Res. 44 (2005) 3894–3900.
[21] Y. Yang, Y. Liu, H. Xiang, Y. Xu, Y. Li, Study on methanolic depolymerization of PET with supercritical methanol for chemical recycling, Polymer Degradradation and Stability 75 (1) (2002) 185–191.
[22] Q. Liu, R. Li, T. Fang, Investigating and modeling PET methanolysis under supercritical conditions by response surface methodology approach, Chemical Engineering J. 270 (2015) 535–541.
[23] A. Schoengen, G. Schreiber, H. Schroeder, 1981. Process for preparation of terephthalic acid by the hydrolysis of intermediate stage crude dimethyl terephthalate. *US Patent 4 302 595.*
[24] M. Sim, M. Han, Hydrolysis of dimethyl terephthalate for the production of terphthalic acid, J. Chem. Eng. Jpn 39 (3) (2006) 327–333.
[25] B. Kim, H. Hwang, D. Woo, M. Han, Design and control of a reactive distillation column based on a nonlinear wave propagation theory: production of terephthalic acid, Ind. Eng. Chem. Res. 49 (9) (2010) 4297–4307.
[26] R.R. Hepner, R.E. Michel, R.E. Trotter, 1995. Process for the separation of glycols from dimethyl terephthalate. US Patent 5 391 263.
[27] S.D. Hall, R.R. Hepner, R.E. Michel, D.R. Wheatcraft Jr., G.M. Williamson, 1999. Process for depolymerizing polyester. US Patent 5 912 275.
[28] G.E. Brown Jr, R.C. O'Brien, 1976. Method for recovering terephthalic acid and ethylene glycol from polyester materials. US Patent 3 952 175.
[29] S.F. Pusztaszeri, 1982. Method for recovery of terephthalic acid from polyester scrap. US Patent 4,355,175.
[30] N.D. Sharma, A.A. Vaidya, P. Sharma, 1985. Recovery of pure terephthalic acid from polyester materials. Ind. Patent 163 385.
[31] T. Yoshioka, T. Sato, A. Okuwaki, Hydrolysis of waste PET by sulfuric acid at 150 °C for a chemical recycling, J. Appl. Polym. Sci. 52 (1994) 1353.
[32] T. Yoshioka, N. Okayama, A. Okuwaki, 1995. Chemical recycling of post consumer PET powder by hydrolysis and simultaneous oxidation in nitric acid. 3rd Int. Symp. East Asian Resources Recycling Tech., Nov. 21–24.
[33] J. Benzaria, B. Durif-Varambon, F. Dawans, J.B. Gailiard, 1994. Process for recovery of alkali metal or alkali earth metal terephtalate and alkylene glycol from alkylene polyterephtalates. Eur. Patent 597 751.

[34] J. Pitat, V. Holcik, M.A. Bacak, 1959. A method od processing waste of polyethylene terephthalate by hydrolysis. GB Patent 822 834.

[35] D. Paszun, T. Spychaj, Chemical recycling of poly(ethylene terephthalate), Ind. Eng. Chem. Res. 36 (1997) 1373.

[36] M.J. Collins, S.H. Zeronian, The molecular weight distribution and oligomers of sodium hydroxide hydrolyzed poly(ethylene terephthalate), J. Appl. Polym. Sci. 45 (1992) 797.

[37] L.-C. Hu, A. Oku, E. Yamada, K. Tomari, Alkari-decomposition of poly (ethylene terephthalate) in mixed media of nonaqueous alcohol and ether. Study on recycling of poly(ethylene terephthate), Polymer J. 29 (9) (1997) 708.

[38] G.P. Karayannidis, A.P. Chatziavgoustis, D.S. Achilias, Poly (ethylene terephthalate) recycling and recovery of pure terephthalic acid by alkaline hydrolysis, Adv Polym Technol 21 (2002) 250.

[39] J.W. Mandoki, 1986. Depolymerization of condensation polymer. U.S. Patent 4 605 762

[40] J.R. Campanelli, D.G. Cooper, M.R. Kamal, Catalyzed hydrolysis of polyethylene terephthalate melts, J. Appl. Polym. Sci. 53 (1994) 985.

[41] A. Michalski, hydrolysis of poly(ethylene terephthalate) waste to obtain terephthalic acid, Wl. Chem. 49 (1987) 144 (in Polish).

[42] A. Michalski, 1987b. Purification of terephthalic acid obtained by hydrolysis of waste PET. Polish Patent 140015.

[43] G. Xi, M. Lu, C. Sun, Study on depolymerization of waste polyethylene terephthalate into monomer of bis(2-hydroxyethyl terephthalate), Polymer Degradation and Stability 87 (2005) 117–120.

[44] F. Pardal, G. Tersac, Kinetics of poly(ethylene terephthalate) glycolysis by diethylene glycol. Part I: Evolution of liquid and solid phases, Polymer Degradation and Stability 91 (2006) 2840–2847.

[45] F. Pardal, G. Tersac, Kinetics of poly(ethylene terephthalate) glycolysis by diethylene glycol. Part II: Effect of temperature, catalyst and polymer morphology, Polymer Degradation and Stability 92 (2007) 611–616.

[46] M.E. Viana, A. Riul, G.M. Carvalho, A.F. Rubira, E.C. Muniz, Chemical recycling of PET by catalyzed glycolysis: kinetics of the heterogeneous reaction, Chemical Engineering J. 173 (2011) 210–219.

[47] A. Sangalang, L. Bartolome, D.H. Kim, Generalized kinetic analysis of heterogeneous PET glycolysis: nucleation-controlled depolymerization, Polymer Degradation and Stability 115 (2015) 45–53.

[48] N. George, T. Kurian, Recent developments in the chemical recycling of postconsumer poly(ethylene terephthalate) waste, Ind. Eng. Chem. Res. 53 (2014) 14185–14198.

[49] H. Kurokawa, M.A. Ohsima, K. Sugiyama, H. Miura, Methanolysis of polyethylene terephthalate (PET) in the presence of aluminum tiisopropoxide catalyst to form dimethyl terephthalate and ethylene glycol, Polymer. Degrad. Stab. 79 (2003) 529–533.

[50] H. Hwang, M. Han, 2010. Kinetics of depolymerization of PET, MS thesis Chungnam National University.
[51] T. Yoshioka, N. Okayama, A. Okuwaki, Kinetics of hydrolysis of PET powder in nitric acid by a modified shrinking-core model, Ind. Eng. Chem. Res. 37 (1998) 336–340.
[52] T. Yoshioka, T. Motoki, A. Okuwaki, Kinetics of hydrolysis of PET powder in sulfuric acid by a modified shrinking-core model, Ind. Eng. Chem. Res. 40 (1998) 75–79.
[53] D.D. Cornell, Modern Polyesters: Chemistry and Technology of Polyesters and Copolyesters, John Wiley & Sons Ltd, 2003 (chapter 16).
[54] U.K. Thiele, Chemical polyester recycling: challenges and hurdles to enter industrial scale, in: 5th China Int'l Recycled Polyester Fiber market & Tech forum, 2–4 September 2009.
[55] M. Genta, F. Yano, Y. Kondo, W. Matsubara, S. Oomoto, 2003. Development of chemical recycling process for post-consumer PET bottle by methanolysis in supercritical methanol. Mitsubishi Heavy Industries Ltd., Technical review, Tokyo, 40, Extra No. 1, pp. 1–4.
[56] M. Genta, R. Uehara, F. Yano, Y. Kondo, W. Matsubara, 2003. Development of chemical recycling process for post-consumer PET bottle by methanolysis in supercritical methanol, in: Proceedings of Sixth ISSF, pp. 1381–1386.
[57] M. Han, 2011. Feedstock recycling process from polyester wastes and apparatus using the same. World Patent (WO) 043 515.
[58] B. Kim, D. Kim, Y. Cho, M. Han, Chemical recycling of poly(Ethylene Terephthalate) using a new hybrid process, J. Chem. Eng. Jpn 41 (9) (2008) 923–928.

6 Chemical Depolymerization of PET Bottles via Ammonolysis and Aminolysis

Prashant Gupta and Subhendu Bhandari
Maharashtra Institute of Technology, Aurangabad, Maharashtra, India

6.1 Introduction

Polyethylene terephthalate (PET) is a saturated polyester synthesized by the esterification scheme of a dibasic acid, i.e., terephthalic acid (TPA) and a diol, i.e., ethylene glycol (EG). Alternatively, it can also be synthesized by the transesterification scheme involving the reaction of ethylene glycol (EG) with dimethyl terephthalate (DMT). Due to the versatility in production of PET over different range of molecular weight in the single multiproduct polymerization process plant and excellent tensile strength, clarity, chemical resistance, processability, and thermal stability, it is used in fibers, bottles, films, and other molded products. The global production of fibers and bottle grades account to be around 60% and 30%, respectively. The existence of PET is both in amorphous as well as crystalline form helps in its availability both as transparent and opaque/white [1–3]. PET is recycled by both physical and chemical routes. The physical route generally consists of remelting (fully/partially) and formation of new products with the help of suitable additives. Furthermore, PET is a condensation product, thereby making it susceptible to solvolytic chain scission. It is chemically reprocessed by total depolymerization into oligomers and other by-products [4]. The extracted monomers are further polymerized to be formed into a new product upon subsequent processing. The regenerated monomer/polymer may be used in the form of blends with pristine materials. The revitalization process in addition to washing, also consist of a lot of polymer refining process such as crystallization, distillation, and other chemical reactions [5]. In addition to washing processes, other additional chemical reactions and processes such as crystallization and distillation may be involved in the regeneration process for the purification of monomer/polymer [6].

Recycling of Polyethylene Terephthalate Bottles.
DOI: https://doi.org/10.1016/B978-0-12-811361-5.00006-7

Chemical recycling is a sustainable form of recycling as it leads to reproduction of PET feedstock. The general chemical methods of PET recycling with the degrading agents consist of glycolysis (glycol), methanolysis (methanol), and hydrolysis (alkalis, mineral acids, water, or steam) to yield monomers, i.e., DMT, EG, TPA, and bis(2-hydroxyethyl)terephthalate (BHET) [7]. The focus of research being carried out in the advancements in chemical recycling of PET is toward increasing the reaction/monomer yield along with progressive decrease in the reaction time under simpler reaction conditions [8].

Furthermore, apart from degradation of PET, chemical recycling methods such as aminolysis also have the potential to bring about surface reactivity changes in PET fibers thereby enhancing its chemical reactivity for applications such as milk protein grafted [9] or carbohydrate grafted [10] PET fabric to enhance the wearing comfort of the fabric.

6.2 Aminolysis

Aminolysis of PET deals with reaction of recycled PET with a host of primary amine solutions such as methylamine, ethylamine, ethanolamine (EA), allylamine, and hydrazine [11–15] in aqueous form correspondingly leading to formation of diamides of TPA and EG. The effect of amines (aminolysis) which are organic bases over polymer degradation is faster than effect of alcohols (i.e., glycolysis). However, utilization of this process on an industrial scale is not reported and is a disadvantage when economy of the process is concerned. The partial aminolysis process has been applied for property enhancement in the production of PET fibers. It is reported that the polymer is taken either in the form of powder or fibers. The temperature range is in the range of 20–100°C. Primary amines were taken along with anhydrous *n*-butyl amine as aminolytic agent at lower temperatures of 21°C. The basic nature of an amine relative to water along with stearic hindrances (size) determines the aminolytic rate of degradation of PET [15]. Furthermore, the reported literature is focused on degradation of PET on a selective basis with the help of aqueous solution of methylamine leading to the formation of methyl terephthalamide (TPD). The morphological changes during degradation make us presume that, the effect of primary amine aqueous solutions help the degradation to start rapidly from the amorphous region, and eventually followed by a significantly sluggish attack would happen on the crystalline regions [16].

The use of complex amines such as 1,2-diaminoethane, 1,6-diaminohexane, 3,6-dioxa-1,8-diaminooctane, and 4,9-dioxa-1,12-diaminododecane [10], diethylenetriamine (DETA), triethylenetetramine (TETA), tetraethylenepentamine (TTEPA) [17], 3-aminopropyltriethoxysilane [18,19] have been reported to carry out chemical modification processes by the thermally induced aminolysis which deals with nucleophilic reaction of organic amine with PET ester bonds resulting in polymer chain scission. Furthermore, a simple and relatively new approach of UV light induced aminolysis for the formation of a tertiary amine terminated PET surface by using dimethyl amine has been reported by Yang et al. [20].

Fukushima et al. [21] reported the aminolysis of PET waste with the help of an organic catalyst, 1,5,7-triazabicyclo[4.4.0]dec-5-ene (TBD). The beverage bottle waste was used after processing (water washing, drying and shredding to dimensions of 3 mm) and deployment of different amines and processing conditions leading to the formation of various TPD compounds as summarized in Table 6.1. The melting temperature was analyzed as the on-set temperature of the melting endotherm. The reaction was carried out in Schlenk tube with different stoichiometric amounts of PET flakes, amine, and TBD. The reaction was carried out in a nitrogen environment with different conditions of temperature and time and the products were washed and filtered with different reagents/solvents. The filtered product was vacuum dried yielding the product in the form of white powder with different yield percentages.

Spychaj et al. [22] reported the use of triethylenetetramine, DETA individually as well as in mixtures. Furthermore, they employed the use of mixtures of triethylenetetramine and *p*-phenylenediamine or triethanolamine by solvolysis at temperatures of 200–210°C in a glass reactor with a ratio of amine to repeating polymeric unit of 2:1. The aminolytic reaction substances were tested for use with polyamines as hardener system for epoxy resin. The solvolytic reaction of polymers comprising of heteroatom (A) in the main chain is degraded as shown in Fig. 6.1. The solvolytic agent (B′B″) is a polyamine such as DETA or TETA, which, upon reaction will result in formation of products with lower molecular weight.

The aminolysis degradation products, solid or viscous liquids, upon characterization with NMR spectra ^{1}H and ^{13}C, show that they are combination of symmetrical primary amide derivatives, symmetrical secondary amides and asymmetrical amide, as shown in Fig. 6.2. The mole percentage of aminolysis reaction products of PET with DETA and TETA was in the range of 69 and 47 for symmetrical primary amide

Table 6.1 Use of Various Amines for PET Depolymerization in Presence of 1,5,7-Triazabicyclo[4.4.0]dec-5-ene as an Organocatalyst [21]

Sl. No.	Amine Used for Aminolysis	Reaction Temperature (°C)	Reaction Time (h)	Yield (%)	Product	T_m (°C) [DSC]
1	Benzylamine	150	1	87	*N,N′*-dibenzylterephthalamide	264
2	Aniline	180	24	41	*N,N′*-diphenylterephthalamide	265
3	Ethylene diamine	110	1	89	*N,N′*-bis(2-aminoethyl) terephthalamide	197
4	1,4-diaminobutane	110	1	82	*N,N′*-bis(4-aminobutyl) terephthalamide	217
5	1,6-diaminohexane	110	2	63	*N,N′*-bis(6-aminohexyl) terephthalamide	171
6	*p*-Xylylene diamine	150	1	79	*N,N′*-bis(4-(amino methyl)benzyl) terephthalamide	223
7	*m*-Xylylene diamine	150	1	67	*N,N′*-bis(3-(aminomethyl)benzyl) terephthalamide	128
8	*p*-Phenylenediamine	190	18	72	*N,N′*-bis(4-aminophenyl) terephthalamide	301
9	*m*-Phenylenediamine	190	18	75	*N,N′*-bis(3-aminophenyl) terephthalamide	295
10	2,6-Diaminopyridine	190	18	72	*N,N′*-bis(6-aminopyridin-2-yl) terephthalamide	233
11	4-aminobenzylamine	120	1	70	*N,N′*-bis(4-aminobenzyl) terephthalamide	203

12	Ethanolamine	120	2	93	*N*,*N*′-bis(2-hydroxyethyl) terephthalamide	228
13	3-(dimethylamino)-1-propylamine	120	3	74	*N*,*N*′-bis(3-(dimethylamino)propyl) terephthalamide	173
14	Allylamine	45	4	81	*N*,*N*′-diallylterephthalamide	220
15	Propargylamine	70	135	66	*N*,*N*′-di(prop-2-yn-1-yl) terephthalamide	205
16	Piperidine	100	16	72	1,4-Phenylenebis(piperidin-1-ylmethanone)	205

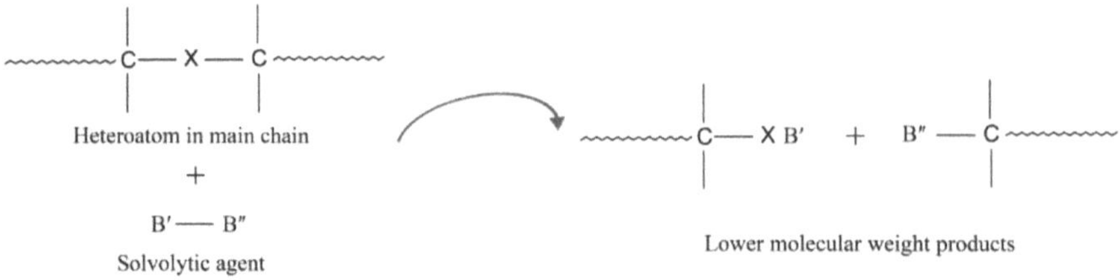

Figure 6.1 Solvolytic degradation reaction of polymers with presence of heteroatoms in the main polymer backbone [22].

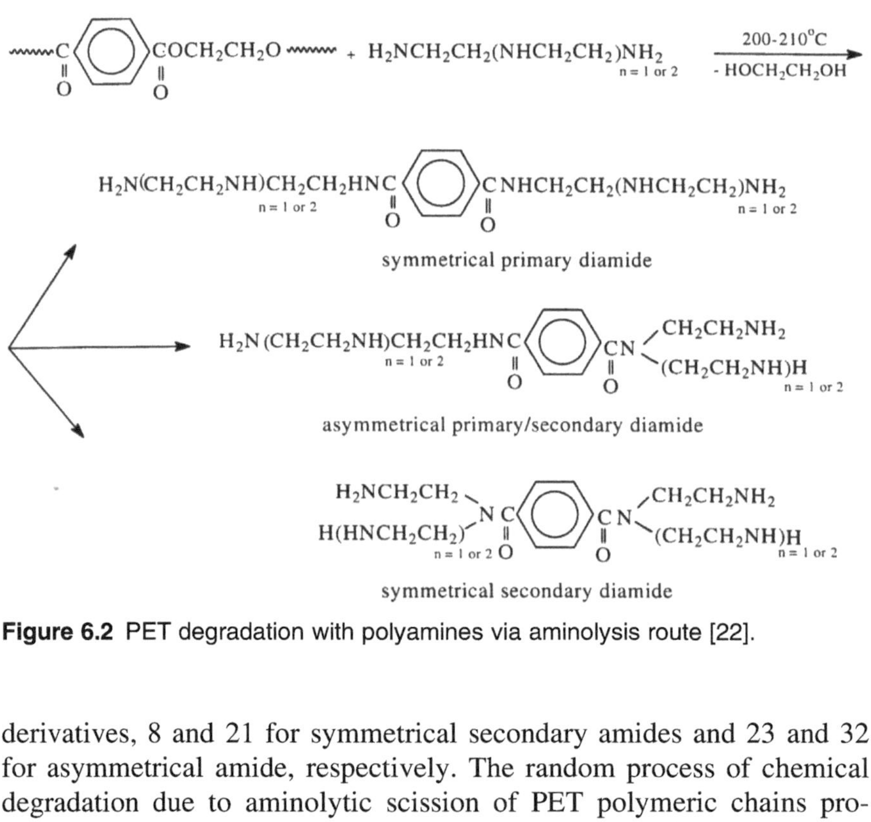

Figure 6.2 PET degradation with polyamines via aminolysis route [22].

derivatives, 8 and 21 for symmetrical secondary amides and 23 and 32 for asymmetrical amide, respectively. The random process of chemical degradation due to aminolytic scission of PET polymeric chains progresses by the attack of primary amino groups on ester bonds present in PET. The degradation products were used in hardening of epoxy resins. They found that hardening of resins was achieved with PET/DETA, PET/TETA, and PET/mixed amine induced degradation products. They exhibited maximum endothermic action at temperatures below 100°C. The usage of aminolysis products of PET based waste material can serve as an important raw material in-feed for reactive resins such as hardening of epoxy resins.

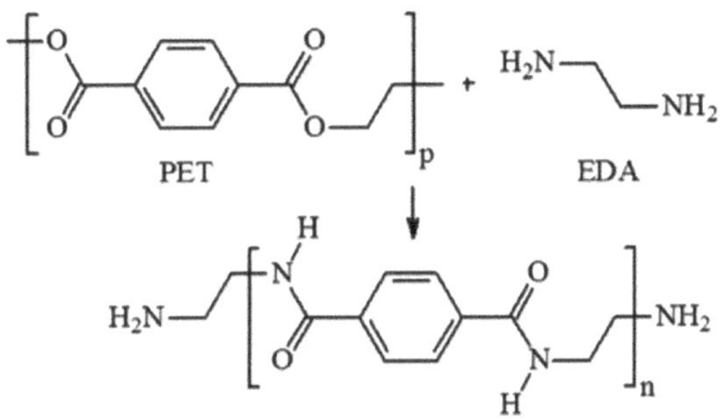

Figure 6.3 The reaction of PET with EDA. $n = 1$: BAET; $n \geq 2$: AOET [23].

Hoang et al. [23] demonstrated aminolysis of PET waste with ethylene diamine (EDA) without the presence of catalyst. PET flakes were reacted with EDA at 100°C for 17 h and further refluxed with addition of methanol for 3 h. The reaction mass (part A) was allowed to cool, rinsed and filtered and dried at 60°C for 24 h in a vacuum oven. The filtrate was further subjected to rotary evaporator for removal of volatiles. The viscous material was left for recrystallization and later filtered, rinsed and dried at 60°C for 24 h in a vacuum oven. The recrystallization of solid mass (part B) was done from dimethyl sulfoxide and it was subjected to further analysis. The reaction scheme is given in Fig. 6.3. EDA reacts with PET leading to forming a half amide-half amino moiety $RCONHCH_2CH_2NH_2$ and hydroxyl end PET. The left amino groups of half amide-half amino moiety is charged to react ahead with another ester group of PET to form higher oligomers such as heptamer, pentamer, etc. These reactions compete with reaction of EDA and ester groups leading to formation of trimer BAET. The acceleration effect of EDA makes it tough to control the reaction using different molar ratios of reactants. However, using protected EDA, although not cost effective can lead to avoiding the same. They confirmed the presence of high molecular weight oligomers of α,ω-aminoligo(ethylene terephthalamide) (AOET) in part A and trimer, bis(2-aminoethyl)terephthalamide (BAET) in part B with the help of analytical testing and evaluation. The FTIR analysis of part A in comparison with part B expresses the disappearance of bands at 3345 and 1579/cm attributed to the primary amine group, and broadening of peak at 3298/cm resembling N−H band due to little contribution of primary amine α,ω-end groups to the oligomeric structure with higher molecular weight. NMR spectra suggest that part A mainly of pentamer whereas part B consisting of trimer BAET. HPLC−MS study suggests that the total ion chromatogram expresses BAET contribution as the main product in part B with 97.3%

contribution with only 2.7% of pentamer product. Part A was insoluble in the reaction mixture and comprises of pentamer (92.5%), i.e., high molecular weight products in its composition as confirmed by mass spectra. They further studied the effect of reactant molar ratios of PET: EDA 1:8 and 1:16 on the oligomer content in part A and found that doubling the concentration of amine increases the pentamer by 4% and decreases the heptamer by 3%.Upon studying the effect of above changes on the yield of BAET and AOET, they found that with an increase in the molar rations of PET:EDA from 1:8 to 1:16, the trimer yield increases by 13% and oligomer yield reduces by 10%. They also suggested the use of amino end caped trimer BAET and oligomers AOET for synthesis of new polyamides and polyimides.

Tawfik and Eskander [24] demonstrated the degradation of PET bottle waste by aminolysis with EA in the presence of dibutyl tin oxide (DBTO) catalyst (Fig. 6.4). The use of EA for aminolysis has an enhanced since it has a hydroxyl group in addition to amine group thereby increasing the potential to cleavage in ester bonds present in polyester. The postconsumer waste was collected, washed, and dried for constant weight. The dried PET waste was depolymerized with EA taken in the molar ration of 1:4 in the presence of catalyst DBTO at 190°C under atmospheric pressure for variable degradation times (1, 2, and 4 h). Addition of distilled water superseded the above step followed by agitation and filtration to remove unreacted amine and oligomeric products. The white filtrate obtained was further purified, in boiling distilled water by formation of small fibers due to evaporation of excess water and chilling. The yielded residue (49%–62%) was dried for further analysis. FTIR analysis evidenced the final white precipitate to be

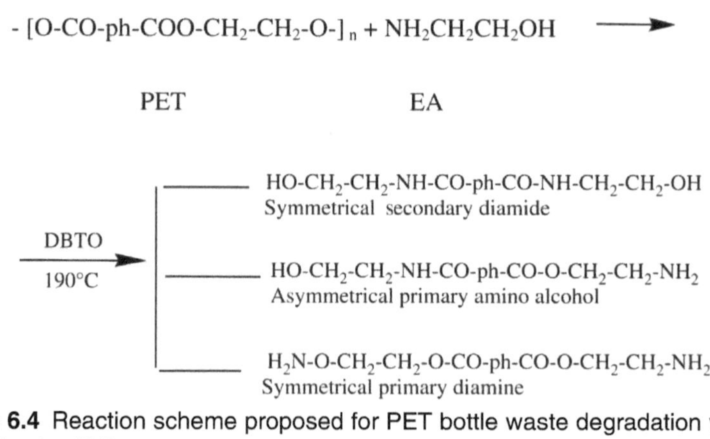

Figure 6.4 Reaction scheme proposed for PET bottle waste degradation with ethanolamine [24].

the same for varying reaction times and catalyst concentration. The products of aminolysis carried out using ethanolamine are highly amorphous which is evident by XRD study and they possess good solubility in DMF, DMS, and hot water. Also, the solubility becomes poor in pyridine, cold water, and acetic anhydride and they are completely insoluble in solvents like butanol, toluene, and ethanol.

Thermal studies indicate that aminolysis products comprising of terephthalate segments have significantly better thermal stability. The onset of T_g is observed at -11°C and crystalline melting point T_m of the residue is observed at 207°C and 234°C which may be due to polydispersity and/or presence of higher oligomers in the degraded product [25]. T_c, i.e., crystallization temperature is observed between T_g and T_m at 217°C. The weight loss upon heating starts from 164°C and further distinguished peaks were observed at 300°C, 369°C, and 610°C for breakdown in hydrogen bonding, polymer chains, and pyrolysis reactions, respectively, during the aminolytic degradation of PET. They concluded the formation of the product identified as bis(2-hydroxyethylene) terephthalamide (BHETA) and proposed that the same can be further used as starting materials for polyurethanes with a potential to be applied in adhesives and coatings.

Shukla and Harad [14] demonstrated aminolysis of postconsumer PET fiber waste as well as bottle waste with EA to form BHETA using different catalysts, e.g., sodium acetate, glacial acetic acid, and potassium sulfate. PET was reacted with EA in the molar ratio of 1:6 using the above-mentioned catalysts at different concentrations up to 8 h under reflux. Addition of distilled water after completion of reaction was to ensure precipitation of the product BHETA. It was further followed by dissolving the precipitate in distilled water for 30 min for concentrating the filtrate and subsequent recrystallization of the product from chilling. BHETA obtained was dried in an oven at 80°C. Shamsi et al. [26] focused on the same process/scheme as shown in Fig. 6.5 using sodium acetate as a catalyst to synthesize polyurethane based on ether-type polyol, 1,4-butanediol and varying molar ratios of hexamethylene diisocyanate.

El-Hameed [27] demonstrated aminolysis PET beverage bottle waste into pol(bis(2-hydroxyethylene)terephthalamide (PHETA) using catalyst and evaluated its application in inhibition of corrosion in carbon steel in presence of acidic environment since they tend to get attacked by acids in applications such as acid cleaning, acid descaling, acid pickling, and oil well acidizing. Aminolysis of PET was carried out with monoethanolamine (MEA) at the ratio of 1:1 in presence of sodium acetate (1 wt.

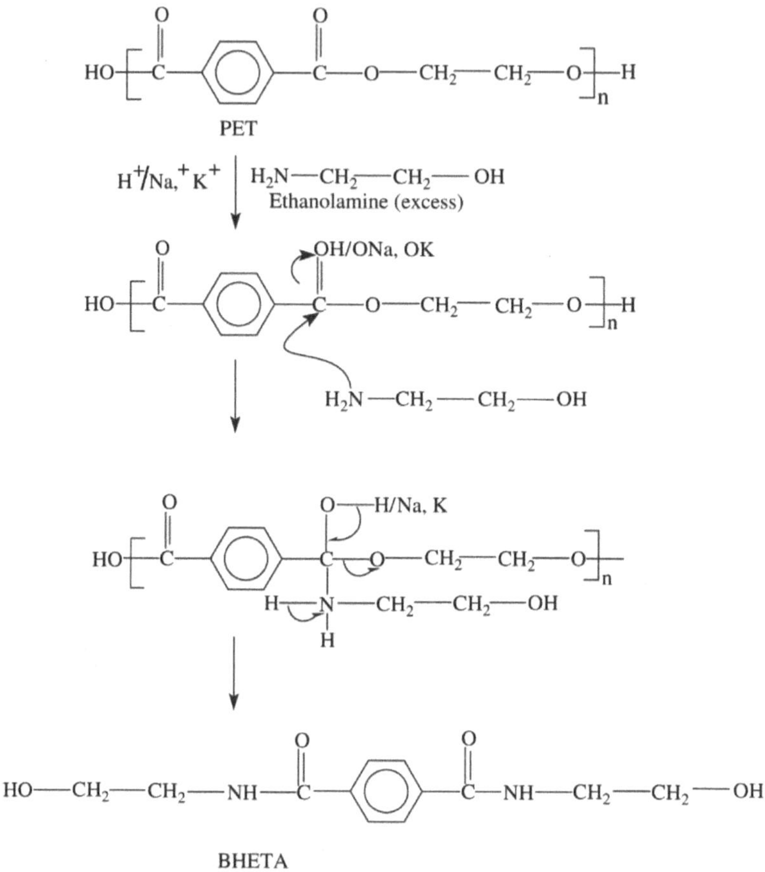

Figure 6.5 Aminolysis of PET with EA to form BHETA [14].

% with respect to PET waste) as a catalyst. The reaction was constantly stirred in inert nitrogen atmosphere for 4 h at 170–190°C followed by 3 h at 200°C. The product was cooled at 100°C and precipitated with distilled water to obtain PHETA which was studied using IR spectra.

The aminolyzed PET product with MEA in the chemical structure increases the water solubility for usage as corrosion inhibitors for carbon steel. PHETA (0–200 ppm) with carbon steel sample and immersed in 1 M HCl for evaluation at different temperatures for corrosion prevention via open circuit potential, potentiodynamic polarization techniques, and weight loss measurements. All the three results suggest that the aminolyzed PET product, i.e., PHETA can function as corrosion inhibitor. The presence of oxygen and nitrogen atoms, which are electronegative, in the form of an electron cloud on the aromatic ring helps in corrosion prevention. The metal surface possesses a negative charge in the acid medium due to adsorbed Cl ions on the surface. The protonation of O_2 and N_2 can easily occur because of high electron density, thereby changing to positively charged inhibitor species. Electrostatic interaction in between inhibitor molecule and oppositely charged metal surface can lead to physical adsorption of inhibitor molecules on the surface. Moreover, formation of coordinate bonds between unshared e-pairs of unprotonated N_2 atoms of the inhibitor and vacant d-orbitals of surface atoms may take place along with chemical absorption of inhibitor molecules due to interaction of π electrons of the benzene ring with vacant d-orbitals of metal [28,29]. The corrosion inhibitor can either be synthesized from pristine raw materials but recycling PET to manufacture the inhibitor can be an economically viable process.

Holmes [30] demonstrated aminolysis of PET fibers with *n*-butyl amine in its aqueous and vapor form. The amine attacking the PET cleaves the polymeric PET chain with one end of the broken chain having an amide group and the other end a hydroxyl group. Aminolysis of PET fibers induces a detrimental change in the characteristics such as molecular weight and tensile strength decrease thereby increasing the degree of crystallinity, with cracks present over the aminolyzed fiber surface.

Furthermore, it renders PET fibers with more wettability, thus enhancing the comfort level of end products such as garments. In addition to that, aminolysis improves the handling and dyeability of PET fibers. The fibers are dipped in aqueous amine solution can be 30%–90% water based and later rinsed in large quantities of water to terminate the degradation reaction and wash away oligomers formed during the same. These huge quantities of water (effluents) are required to be treated before disposal and hence required large amount of energy and thereby is not cost effective. However, it can be saved if the reaction can be carried out in the vapor phase, by exposing the fibers to neat amine vapor in a closed vessel. In this way, whole of the amine can be used for degradation as no chances of oligomer adulteration shall exist. Furthermore, water will not be required during reaction and rinsing is the only stage where it would be employed which shall save huge quantities of effluent generated. The treatment of PET fibers with aqueous and vapor *n*-butylamine was carried out in enclosed jars at 21°C. The aqueous solution comprises of 56% (v/v) *n*-butylamine. The aminolyzed fibers were rinsed in methanol and washed with distilled water until neutralization as indicated by litmus paper. The samples were dried in air and further conditioned at 21°C and 65% relative humidity. The selected samples were extracted with chloroform by using Soxhlet apparatus for a period of 8 h, followed by drying and conditioning to constant weight overnight. It was found that the aminolysis employed in vapor phase had 4% weight loss up to a period of 240 days with M_v loss of 94%. The results of aqueous phase aminolysis were contrasting with 21% weight loss during the same period with M_v loss of 74%. This is attributed to removal of oligomers formed by aminolytic degradation reaction in the aqueous amine medium which is not the case in aqueous amine medium. The aqueous treatment of fibers in *n*-butylamine solution allowed the aminolysis products to assimilate in the solution which could not happen in vapor medium with rinsing removing some of the oligomers formed. This was further concluded with analysis of Soxhlet extracted samples where the weight loss had a dramatic increase in the weight loss values in the vapor phase aminolysis reaction product. The presence of some oligomers was evident even in aqueous phase product sample since there was a slight increase in the same as well. The density measurements indicated increase in the density by 0.43% and 0.14% in vapor and aqueous phase aminolysis. Although these results are not very pronounced, they indicate attack and subsequent removal of the amorphous region in the PET fibers. The tensile strength results, expressed that the breaking load decreased with increase in the exposure time. The vapor aminolyzed samples property deterioration was swifter than aqueous aminolyzed samples. The measured loss was 89% and 56% for vapor and aqueous treatment, respectively, tested for 168 h after which the samples could not be tested since they turned too fragile upon exposure. They explained aminolysis as a selective reaction occurring, to start with, prominently in the unoriented amorphous regions. Furthermore, four stages of aminolysis reaction are given as follows:

1. A reduction in molecular weight with virtue of the attack on the unoriented amorphous regions with little change in weight. Also changes in crystallinity are negligible in stage I.
2. Increased weight loss due to continual attack on amorphous phase/regions.
3. A substantial decrease in tenacity by the attack on oriented amorphous regions and crystalline phase.
4. Relaxation of reaction rate with decelerating attack over both crystalline and amorphous regions.

Scanning electron microscopy revealed surface cracks for vapor and aqueous phase aminolysis. As shown in Fig. 6.6, aminolysis of PET fibers in vapor phase led to axial cracks which are different to those of

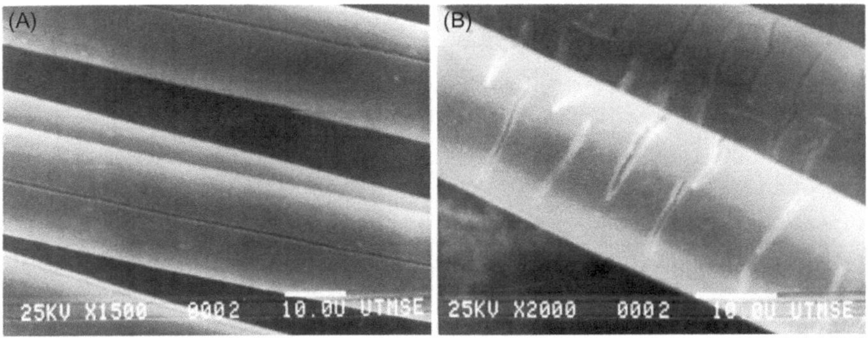

Figure 6.6 SEM micrographs of PET fibers aminolyzed in (A) *n*-butylamine vapor and (B) aqueous *n*-butylamine for 144 h [30].

aqueous phase with radial cracks. The oppositely oriented cracks lead to difference in the direction of attack of amine in vapor versus aqueous phase. The greater crack dimension of aqueous *n*-butylamine aminolyzed samples refer to the hypothesis of attack on the unoriented amorphous region in the direction of fiber radius whereas the smaller one over the extended tie molecules resembles the attack over the oriented fiber in the direction of fiber axis.

Bulak and Acar [31] demonstrated the use of degradation products of PET aminolysis for preparation of paint binder based on alkyd resin. The beverage bottle waste was depolymerized with 1:1 ratio of PET and diethyl amine in presence of zinc acetate catalyst at 220°C for 3 h. The synthesis of alkyd resin was done with soybean oil fatty acid, phthalic anhydride, glycerine, and depolymerization products. The yield of solids was around 98% with Mn of 1516. The comparison of the alkyd resin synthesized with aminolyzed PET was made with standard alkyd resin synthesized using EG. The properties examined during the study such as hardness, adhesion, drying time, abrasion resistance, impact resistance, gloss, chemical resistance (salt water, alkaline, solvent), etc. can help in determining the utility of PET aminolysis product as a substitute of glycol in alkyd resin preparations. They observed that the films made with aminolyzed PET formulation were transparent and homogeneous with no detrimental effect on physical film properties. The uncured resin films were soft and glossy with high adhesion values which turns into glossy, medium hard and relatively elastic with high adhesion, abrasion, and impact strength values after oven curing at 110°C for 1 h.

Shamsi et al. [26] experimented on using aminolysis for development of an eco-friendly technique for PET waste processing at low temperatures and pressure. The aim was to use less time and energy to deplete the PET waste from the environment. The process results into an amorphous product with better thermal stability with a high percentage of yield. The process deals with reaction of PET waste with ethanolamine to obtain BHETA in its pure form for yields in the range of 90% which is unique considering the other decomposition methods such as glycolysis, hydrolysis, etc. need separate steps for purification to be carried out due to the formation of low molecular weight by products, i.e., oligomers. Strong hydrogen bonds are evident from the FTIR spectra for standard PU (with traditional raw materials) as well as BHETA (obtained from aminolyzed PET) based PU. DSC further confirmed cross linking, H bonding, and chain scission with both with and without BHETA-PU samples. The shear strength tests carried out on Fe–Fe and Al–Al substrates exhibit higher maximum load and elongation from BHETA-PU in comparison with standard PU. This indicates that, the polyurethanes synthesized by aminolyzed PET product BHETA, can be further reacted for use in certain segments such as adhesives and polymeric coatings.

In the work of Popescu et al. [32], combination of saponification individually or simultaneously along with aminolysis was found to be beneficial toward improvement of tinctorial response of recycled PET. Increase of amorphous phase after recycling improves dyeability of PET. Generation of amine groups after aminolysis enhances dyeability of PET with anionic dyes with more intensity, whereas, dyeability using cationic dyes is improved by saponification treatment owing to generation of ester groups. Moreover, combination of saponification and aminolysis leads to improvement in dyeability toward nonionic dyes like disperse dye.

6.3 Ammonolysis

Ammonolysis in comparison to other chemical recycling methods has been less explored. The contributing factor toward this can be summarized onto the economic application of degradation products which shall determine the overall feasibility of the degradation process. The degradation process is carried out by the action of ammonia on PET. It is much slower than aminolysis and usually needs the use of a catalyst to increase the rate of degradation.

Mittal et al. [33] used PET bottle waste in the form of flakes washed in hot water at 70–100°C and then with distilled water before drying at 60°C. The reaction of ammonolysis was carried out with aqueous ammonia in a ratio of 1:10 w/v at room temperatures of 40°C with constant agitation. The flakes were separated from the precipitate, washed with distilled water and vacuum dried for further analysis. The depolymerization of PET with the help of anhydrous ammonia produces TPD and EG as per the chemistry of the process shown in Fig. 6.6 with and without catalyst, i.e., quaternary salt. It was observed that TPD was obtained via ammonolysis at varying intervals of time, i.e., 3–45 days indicating the same mechanism being followed over time. The separation of PET waste was possible up to 30 days of reaction time after which brittleness of flakes leads to inability of separation of the same from untreated PET waste. The weight of unreacted PET and degraded product are summarized in Table 6.2. It is evident from Table 6.2 that the presence of a quaternary amine as a catalyst slightly increases the rate of degradation reaction. The rate of degradation is slightly faster with the reaction carried out in the presence of catalyst.

A two-phase degradation process which occurs by nucleophilic mechanism resulting in slow nature of the reaction. Electrophilic mechanism is much faster but since ammonia is a weak acid than water and such acids hardly catalyze the electrophilic mechanism. The obtained product was white solid, i.e., 100% pure TPDs, to conclude whether it has happened due to random chain scission as the product obtained was only one which is not consistent with random chain scission which generally may have other TPDs with the product. If we consider unzipping reaction, once again it starts from one end and completes at other which is once again not consistent with PET as it is a semicrystalline product wherein amorphous region will get degraded faster than the crystalline counterpart and hence the simultaneous action of both these reactions can be understood as the mechanism of degradation process (Fig. 6.7).

Table 6.2 Change of Weight (%) of Unreacted PET and Degraded Product During Ammonolysis [33]

No of Days	% Weight Change of Unreacted PET Waste		% Weight Change of Degraded Product	
	With Catalyst	Without Catalyst	With Catalyst	Without Catalyst
0	100	100	0	0
3	90	94	7	6
7	81	86	17	14
15	69	73	30	27
21	62	71	36	29
25	60	69	38	31
30	PET flakes becomes brittle and difficult to separate			
45	PET flakes becomes brittle and difficult to separate			

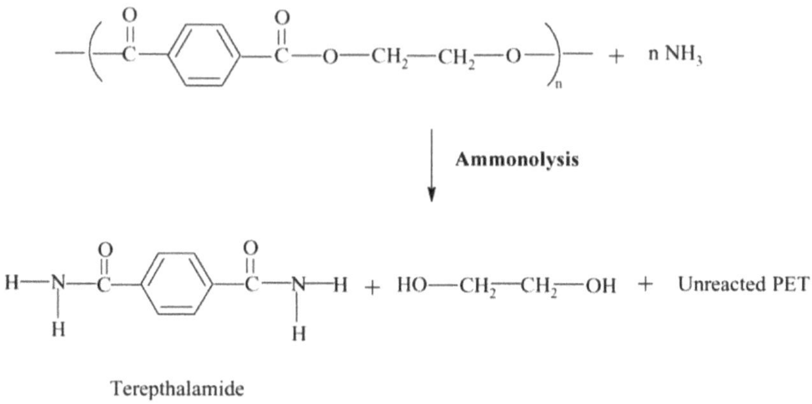

Figure 6.7 Mechanism of ammonolytic cleavage of PET [33,34].

Scanning electron micrographs show that the bulk of PET waste is getting converted into degraded products as shown in Figs. 6.8 and 6.9. Fig. 6.8A shows the SEM micrographs of waste PET flakes before starting ammonolysis degradation reactions wherein smooth, continuous, and flaky structure can be observed. The semicrystalline structure with more crystallinity can be observed in the micrograph with scattered amorphous regions between the oriented crystalline ones. An apparent increase in degradation is observed in SEM micrographs shown in Fig. 6.8B and C which are for 8 days of degraded sample with and without catalyst, respectively. Fig. 6.8D–G shows progressive degradation upon 21 and 30 days of continuous ammonolysis reactions with formation of more and more discontinuous structures. A look at the comparative graphs provides further evidence of progression of degradation as nearly the same irrespective of the presence of a catalyst.

The SEM micrographs, Fig. 6.9A and B, i.e., of short brittle crushable flakes in the form of product obtained after ammonolytic degradation of PET show the further progress of degradation reaction with and without the presence of catalyst after 45 days. The degradative attack of ammonia on both amorphous and crystalline regions of PET simultaneously, however the selectivity to separate them is low when compared with other such reactions such as aminolysis in the presence of methylamine. It might be due to the dependency of the selectivity of raw materials based on its nucleophilicity. Methylamine can selectively separate crystalline and amorphous phases of PET due to high nucleophilicity in comparison with ammonia which brings about ammonolysis degradation reactions in PET. The above investigations determine that the long PET polymeric chains present in the semicrystalline polymer are reduced to

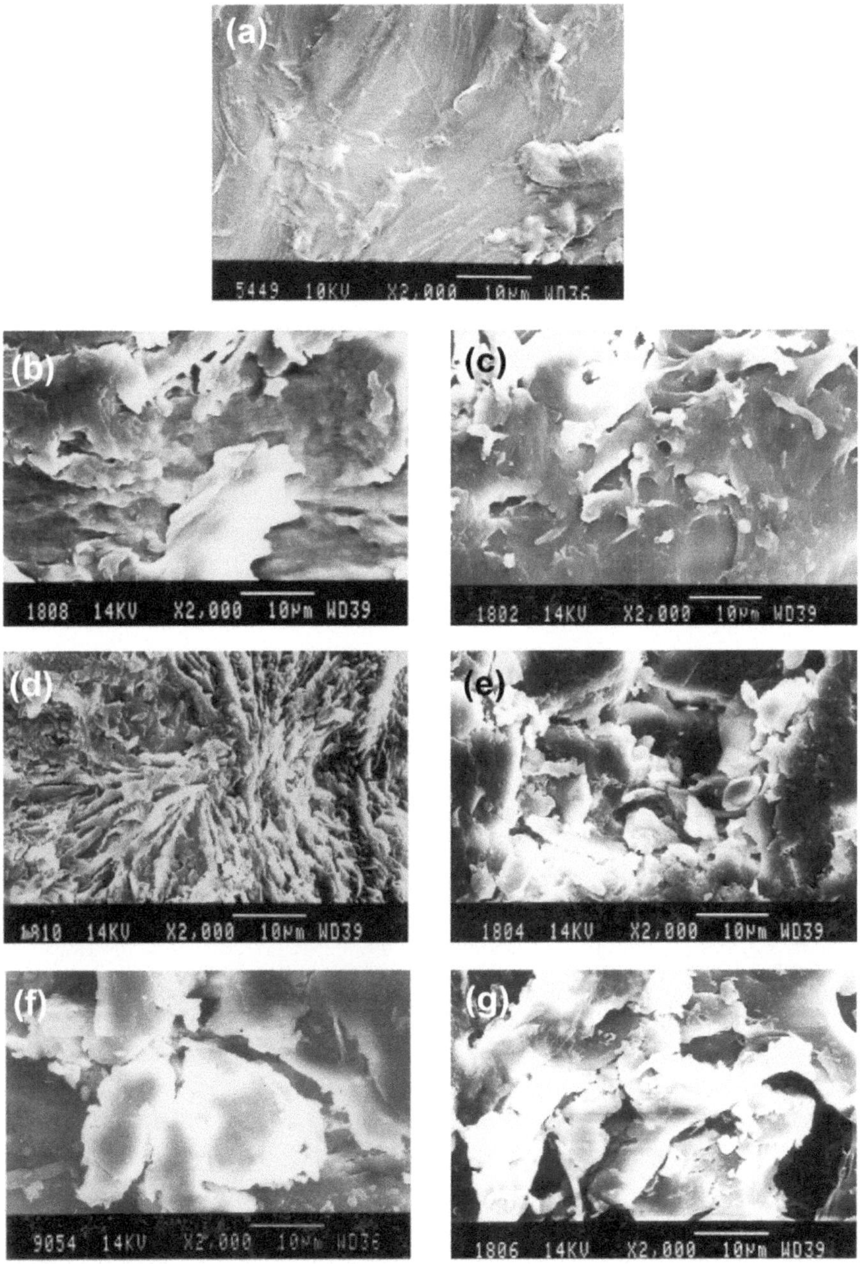

Figure 6.8 SEM of PET flakes (A) without degradation (fresh sample), (B) after 7 days with catalyst, (C) after 7 days without catalyst, (D) after 21 days with catalyst, (E) after 21 days without catalyst, (F) after 30 days with catalyst, (G) after 30 days without catalyst [34].

monodisperse rods and fissures are observed on the surface which has increased in depth with time due to the action of degradation from the surface travelling toward the bulk of PET.

PET is converted into TPD with the action of concentrated anhydrous ammonia in an environment of EG. Terephthalamide serves as a feedstock for manufacture of value added products as it is not a very economically important chemical in the same chemical form. Fig. 6.10 illustrates the chemistry involved in synthesizing of TPD derivatives.

A US patent filed by Blackmon et al. [36] from General Electric for the ammonolysis of PET waste to amine functional monomers such as 4-aminomethyl benzoic acid, *p*-xylylene diamine and 1,4-bis(aminomethyl)cyclohexane which can be employed as feedstocks for polyamides. The three-step process involved in the above conversion starts with TPD synthesis by aminolysis of PET followed by pyrolytic dehydration of the TPD into terephthalonitrile. Lastly, obtained terephthalonitrile is hydrogenated to produce diamine monomer, i.e., *p*-xylylene diamine and/or *p*-xylylene diamine (1,4-bis(aminomethyl)cyclohexane).

Figure 6.9 SEM of the (A) SEM of the product obtained from ammonolysis of PET flakes with catalyst. (B) Product obtained from ammonolysis of PET flakes without catalyst [34].

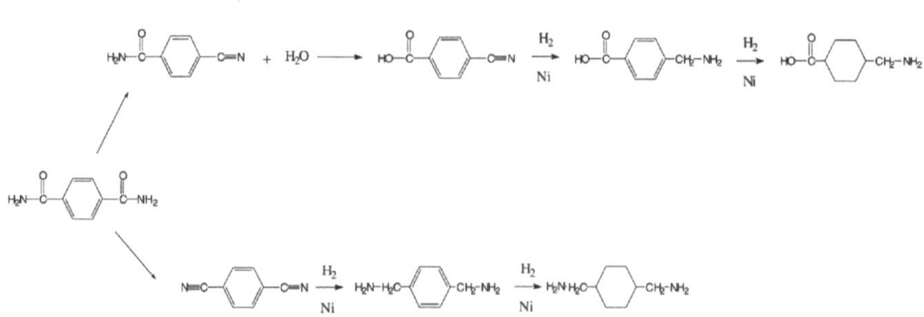

Figure 6.10 Formation of TPD derivatives [35].

As shown in the process outline in Fig. 6.11, PET waste is dispersed in EG. The partial dissolution of PET in EG serves as a medium for heat transfer and reaction media. Thereafter, ammonia is added to the aqueous solution for bringing upon its contact with PET. The application of heat and pressure brings about the reaction wherein PET is degraded upon reaction with ammonia to convert to TPD and EG. The reaction mixture is heated within the temperature range of 120–180°C and pressures of 125 psi yielding more than 90% yield with purity in excess of 99%. In general, ammonolysis reactions may be carried out at temperatures of 80–250°C. Excessively lower temperatures yield slow reactions whereas higher temperatures than above lead to quick side reactions which are undesirable. The product obtained is insoluble in EG and separation is done by filtration to yield TPD in powdered form. It is then washed with deionized water, filtering, and drying at 80°C for 1 h to obtain pure TPD. Pyrolytic degradation of TPD involves heating the obtained TPD to pyrolysis temperature of 340–370°C. At temperatures below 300°C, the reactions are sluggish whereas undesirable side

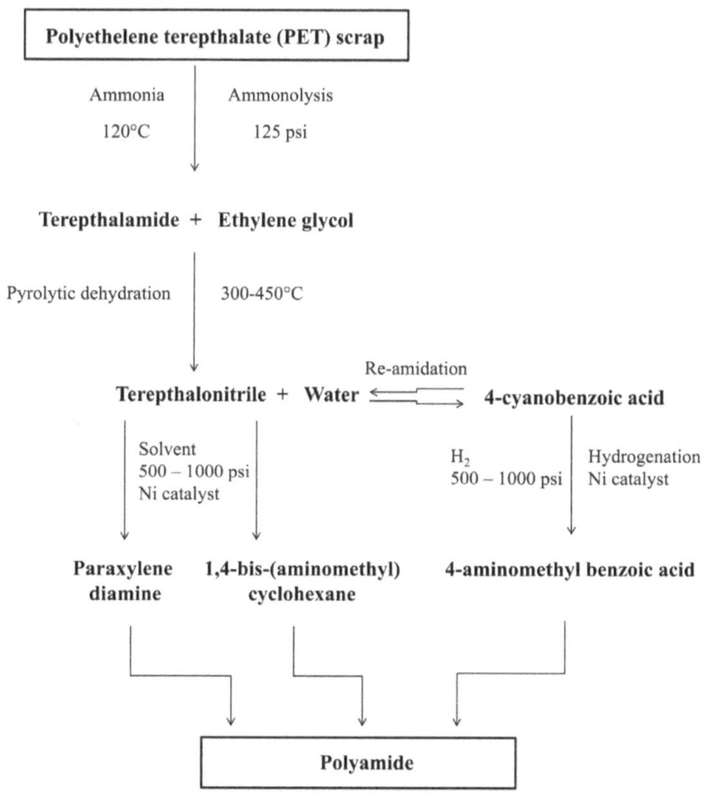

Figure 6.11 Flow chart for the ammonolysis of PET and formation of products.

reactions occur too rapidly at temperature above 450°C. The reactions are carried out in diatomic oxygen free environments avoiding oxidative reactions to produce substantial yield of terephthalonitrile which is separated by addition of hot ethanol to dissolve terephthalonitrile and precipitating insoluble TPD further separated by filtration. Solvent evaporation method may be used for separation of terephthalonitrile. The obtained product can be further processed by hydrogenation to yield useful diamines. There are chances of partial and side reactions during the pyrolytic dehydration of TPD. The dehydration process occurs in two stages wherein the first stage deals with the loss of one molecule of water producing 4-cyanobenzamide which can be converted into terephthalonitrile by further addition of heat. The resulting product is subjected to pyrolytic dehydration to yield terephthalonitrile.

The formation of by-products as mentioned above during the pyrolytic degradation by side reactions can be illustrated by the action of water of the main reaction with partial dehydration product to form 4-aminobenzoic acid. It can be either recycled by reamidation to increase the yield of terephthalonitrile or hydrogenated (reduced) to form 4-aminomethyl benzoic acid, a bifunctional monomer used for synthesis of polyamides. The terephthalonitrile is hydrogenated to produce *p*-xylene diamine and/or 1,4-bis(aminomethyl)cyclohexane. A blend of 90:10 ratio of ethanol to ammonia is used for the process of dehydrogenation in presence of nickel catalyst, Ni–5136P with 65% Ni in its composition on a Ni–silica support are used in combination with pressurized hydrogen at 500–1000 psi and temperature of around 100–200°C for reaction times of 2–8 h. Mild conditions favor the formations xylylene diamines with more stringent conditions favoring formation of cyclohexane derivatives.

Jain and Soni [37] demonstrated a room temperature low-pressure method with the use of zinc acetate as catalyst to obtain TPD or 1,4-benzene dicarboxamide, an aromatic amide which can be further used as hardeners in epoxies. PET flakes were taken and allowed to mix in 10 times w/w solution of ammonia separately with and without the presence of 1% w/w zinc acetate in sealed reaction flasks at ambient temperature and pressure. The precipitates start after hours of mixing and the reaction is carried out at time intervals of 3, 7, 15, 21, 25, 30, and 45 days. They are recrystallized with ethanol, washed with distilled water followed by drying under vacuum for 2 h at 60°C. The presence of catalyst accelerated the degradation process and PET flakes were not observed after 15 days of constant stirring of the reaction mixture which was only obtained after 45 days of stirring without catalyst in the reaction mixture.

Fig. 6.12 shows the FTIR analysis of indicate the formation of ammonolysis products in the presence of catalyst. After 3 days of reaction in the presence of catalyst with peaks at 1750–1730/cm and 1100–1000/cm, resembling aromatic ester C = O str. and aromatic ester C–C–O str. in unreacted PET. However, these peaks are not visible in the IR spectra of the white colored precipitate obtained after 15 days (Fig. 6.12C), thereby indicating the conversion of PET flakes into TPD within 15 days in the presence of zinc acetate. Mass spectroscopy evaluation also further provides supportive evidence of ammonolytic degradation of PET. A molecular ion peak at *m/z* 148 along with fragmented peaks at *m/z* 120, 104, and 76 which could be from TPD is visible in the mass spectra of white precipitated product.

Scanning electron micrographs as shown in Fig. 6.13 of the ammonolyzed samples after 15 days of degradation in the presence of catalyst reflect the complete degradation of PET flakes into TPD. The samples with 15 days of ammonolytic degradation in absence of catalyst show relatively smooth and continuous nondegraded structures which looks like the starting sample. The 15-days ammonolyzed sample in the presence of catalyst show degraded structures in the micrographs which resemble the 45-days degraded sample without the presence of catalyst. Dynamic thermal analysis (DTA) of flakes before and after degradation shows single endothermic peak at 247.1°C and 333.7°C, respectively, in which the latter resembles the peak for pure TPD which means the conversion of PET to TPD. The action of zinc acetate [$Zn(OAc)_2$] in the reaction mixture enables the exchange of its anion with the nucleophile in excess through the aqueous phase. The attack of $Zn(OAc)^+$ and NH_2^- on the polymer and subsequent displacement leads to the formation of TPD. Thus, the above process serves as an environment friendly process and technology for PET waste recycling under ambient conditions of temperature and pressure which aims at reduction in the amount of PET contribution in solid waste stream, thereby consuming less time and saving energy.

Another low-pressure process has been reported by Sharma et al. [38] wherein ammonolysis of PET is brought about by ammonia in EG environment by employing 0.05% w/w zinc acetate in the reaction mixture at temperature of around 70°C with PET:ammonia ratio of 1:6. The degradation reaction yielded 87% TPD.

6.4 Conclusion

The focus is always on sustainable development and chemical recycling certainly has its part to play due to high consumption of PET in

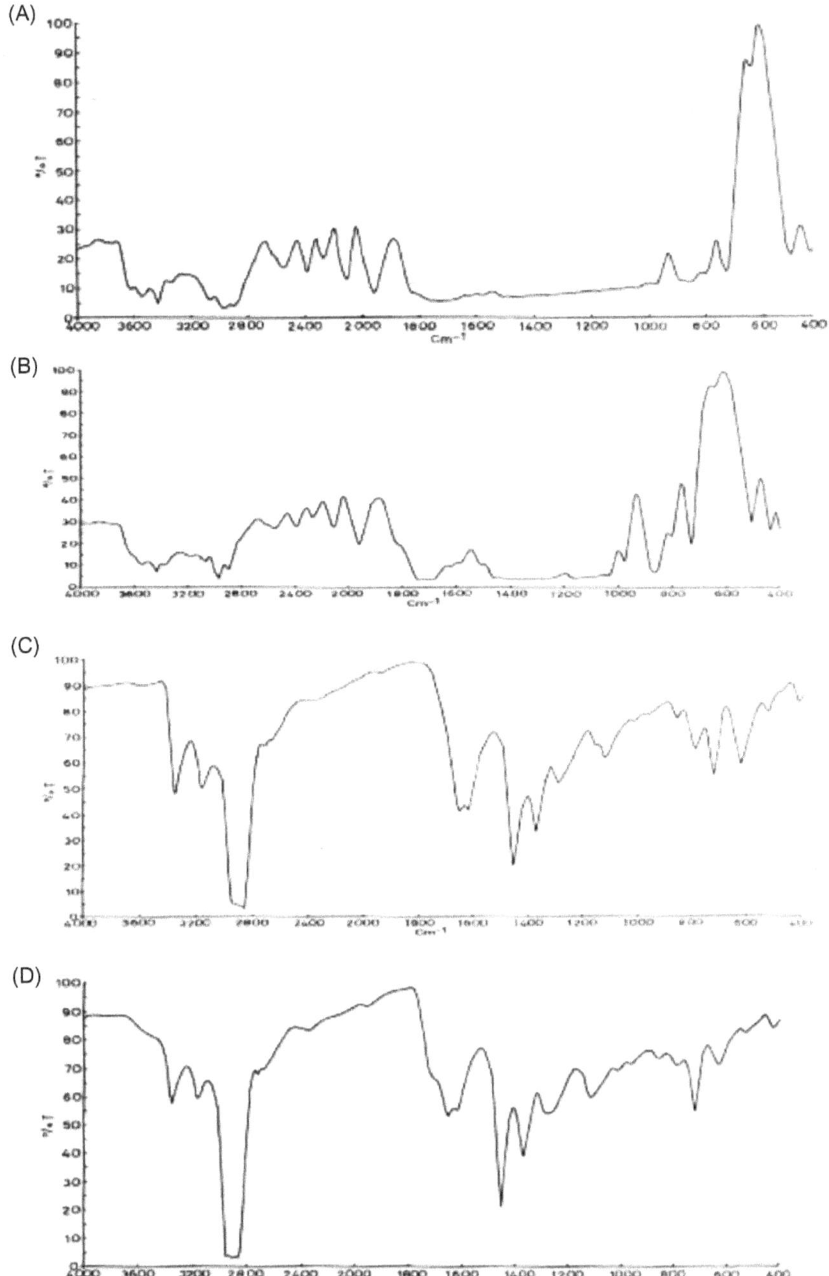

Figure 6.12 IR spectra of (A) ammonolyzed product after 3 days, (B) after 7 days, (C) after 15 days in presence of catalyst, and (D) reference sample [37].

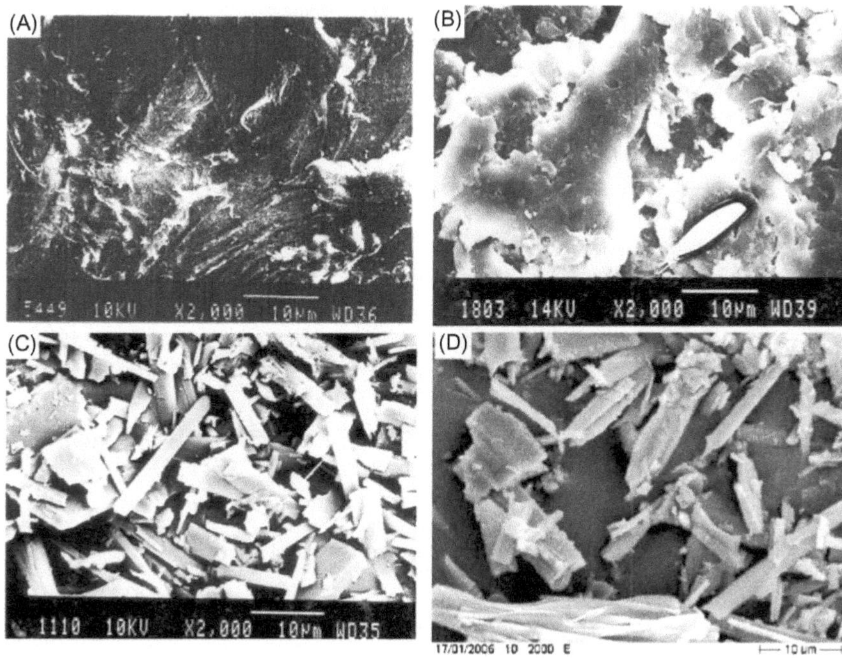

Figure 6.13 SEM of PET flakes sample (A) before degradation, (B) after 15 days of degradation without catalyst, (C) after 45 days of degradation without catalyst, and (D) after 15 days of degradation with catalyst [37].

today's world. The chemical recycling of PET through aminolysis and ammonolysis poses its own economic challenges. The process must be technologically modified to be eco-friendly in terms of either providing good quality substitute raw materials at a cheaper cost or to reduce the time and energy consumption of the manufacturing process. There are some PET chemical recycling units in Japan as they show high interest in recycling and thereby rationalizing the use of nonrenewable resources. The discontinuity of supply along with insufficient homogeneity of PET waste supplies are the challenges due to which the continuous processes have not been as successful commercially as a project may demand. The yield, quality of raw materials, and economics of the process have limited the exploration of both these chemical recycling methods commercially and more so of ammonolysis even on the R&D scale. There is a huge scope for ammonolysis to be used for production of polyamides like poly *p*-phenylene TPD which has a good commercial value. Also, other products such as hardeners for epoxy resin, polyurethane binders and paints, and various other polyamides can be manufactured using them.

References

[1] V.B. Gupta, Z. Bashir, PET fibres, films, and bottles, in: S. Fakirov (Ed.), Handbook of Thermoplastic Polyesters, Wiley-VCH, Weinheim, Germany, 2002, pp. 317–388.

[2] V. Jankauskaite, G. Macijauskas, R. Lygaitis, Polyethylene terephthalate waste recycling and application possibilities: a review, Mater. Sci. (Mediziagotyra) 14 (2) (2008) 119–127.

[3] C.A. Miclău, M. Ienaşcu, I. Ursu, D.C. Tănasie, M.N. Stefănuţ, Chemical recycling of polyethylene terepthalate (PET) waste using sub and supercritical water, Revue Roumaine de Chimie 60 (5–6) (2015) 579–585.

[4] A. Firas, P. Dumitru, Recycling of PET, Eur. Polym. J. 41 (7) (2005) 1453–1698.

[5] D.E. Nikles, M.S. Farahat, New motivation for the depolymerization products derived from poly(ethylene terephthalate) (PET) waste: a review, Macromol. Mater. Eng. 290 (1) (2005) 13–30.

[6] J.R. Overton, S.K. Haynes, Determination of the crystalline fold period in poly(ethylene terephthalate), J. Polym. Sci.: Polym. Symp. 45 (1) (1973) 9–17.

[7] A.H. Haghighi, Y. Sefidbakht, K. Shekoohi, A. Ghaderian, M. Khoonkari, Chemical recycling of PET wastes with different catalysts, Int. J. Polym. Sci. (2015) 1–11. Available from: https://doi.org/10.1155/2015/124524. Article ID 124524.

[8] L. Bartolome, M. Imran, B.G. Cho, W.A. Al-Masry, D.H. Kim, in: D. Achilias (Ed.), Recent developments in the chemical recycling of PET in material recycling-trends and perspectives., InTech, 2012. Available from: https://doi.org/10.5772/33800.

[9] J.F. Zhou, D.D. Zheng, L. Zhong, F.X. Zhang, G.X. Zhang, A novel preparation of milk protein/polyethylene terephthalate fabric, IOP Conf. Series: Mater. Sci. Eng. 137 (2016) 012050. Available from: https://doi.org/10.1088/1757-899X/137/1/012050.

[10] L. Bech, T. Meylheuc, B. Lepoittevin, P. Roger, Chemical surface modification of poly(ethylene terephthalate) fibers by aminolysis and grafting of carbohydrates, J. Polym. Sci. A: Polym. Chem. 45 (11) (2007) 2172–2183.

[11] Y.W. Awodhi, A. Johnson, R.A. Peters, P.V. Poopola, The aminolysis of poly(ethylene terephthalate), J. Appl. Polym. Sci. 33 (7) (1987) 2503–2512.

[12] V.A. Popoola, Polyester formation: aminolytic degradation and proposed mechanisms of the reaction, J. Appl. Polym. Sci. 36 (7) (1988) 1677–1683.

[13] M.S. Ellison, L.D. Fishcr, K.W. Alger, S.H. Zeronian, Physical properties of polyester fibers degraded by aminolysis and by alkaline hydrolysis, J. Appl. Polym. Sci. 27 (1) (1982) 247–257.

[14] S.R. Shukla, A.M. Harad, Aminolysis of polyethylene terephthalate waste, Polym. Degrad. Stab. 91 (8) (2006) 1850–1854.
[15] M.J. Collins, S.H. Zeronian, M.L. Marshall, Analysis of the molecular weight distributions of aminolyzed poly(ethylene terephthalate) by using gel permeation chromatography, J. Macromol. Sci. A: Chem. 28 (8) (1991) 775–792.
[16] C. Farrow, D.A.S. Ravens, I.M. Ward, The degradation of polyethylene terephthalate by methylamine—study by infra-red and X-ray methods, Polymer 3 (1962) 17–25.
[17] Y. Avny, L. Rebenfeld, Chemical modification of polyester fiber surfaces by amination reactions with multifunctional amines, The Analyst 32 (3) (1986) 4009–4025.
[18] L.N. Bui, M. Thompson, N.B. McKeown, A.D. Romaschin, P.G. Kalman, Surface modification of the biomedical polymer poly(ethylene terephthalate), The Analyst 118 (5) (1993) 463–474.
[19] A.Y. Fadeev, T.J. McCarthy, Surface modification of poly(ethylene terephthalate) to prepare surfaces with silica-like reactivity, Langmuir 14 (19) (1998) 5586–5593.
[20] P. Yang, X. Zhang, B. Yang, H. Zhao, J. Chen, W. Yang, Facile preparation of a patterned, aminated polymer surface by UV-light-induced surface aminolysis, Adv. Funct. Mater. 15 (9) (2015) 1415–1425.
[21] K. Fukushima, J.M. Lecuyer, D.S. Wei, H.W. Horn, G.O. Jones, A. Hamid, et al., Advanced chemical recycling of poly(ethylene terephthalate) through organocatalytic aminolysis, Polym. Chem. 5 (2013) 1610–1616.
[22] T. Spychaj, E. Fabrycy, S. Spychaj, M. Kacperski, Aminolysis and aminoglycolysis of waste poly(ethylene terephthalate), J. Mater. Cycl. Waste Manag. 3 (2001) 24–31.
[23] C.N. Hoang, Y.H. Dang, Aminolysis of poly(ethylene terephthalate) waste with ethylenediamine and characterization of α,ω-diamine products, Polym. Degrad. Stab. 98 (3) (2013) 697–708.
[24] M.E. Tawfik, S.B. Eskander, Chemical recycling of poly(ethylene terephthalate) waste using ethanolamine, sorting of the end products, Polym. Degrad. Stab. 95 (2) (2010) 187–194.
[25] G. Colomines, A.V.D. Lee, J.J. Robin, B. Boutein, X-ray diffraction of the crystallinity of glycolysates derived from PET, Eur. Polym. J. 44 (9) (2008) 2874–2885.
[26] R. Shamsi, M. Abdouss, G.M.M. Sadeghi, F.A. Taromi, Synthesis and characterization of novel polyurethanes based on aminolysis of poly(ethylene terephthalate) wastes, and evaluation of their thermal and mechanical properties, Polym. Int. 58 (1) (2009) 22–30.
[27] A.R.S. El-Hameed, Aminolysis of polyethylene terephthalate waste as corrosion inhibitor for carbon steel in HCl corrosive medium, Adv. Appl. Sci. Res. 2 (3) (2011) 483–499.

[28] A.K. Singh, M.A. Quraishi, Inhibitive effect of diethylcarbamazine on the corrosion of mild steel in hydrochloric acid, Corros. Sci. 52 (4) (2010) 1529–1535.
[29] R. Solmaz, G. Kardas, M. Culha, B. Yazici, M. Erbil, Investigation of adsorption and inhibitive effect of 2-mercaptothiazoline on corrosion of mild steel in hydrochloric acid media, Electrochim. Acta 53 (20) (2008) 5941–5952.
[30] S.A. Holmes, Aminolysis of poly(ethylene terephthalate) in aqueous amine and amine vapor, J. Appl. Polym. Sci. 61 (2) (1996) 255–260.
[31] E. Bulak, I. Acar, The use of aminolysis, aminoglycolysis, and simultaneous aminolysis–hydrolysis products of waste PET for production of paint binder, Polym. Eng. Sci. 54 (10) (2014) 2272–2281.
[32] V. Popescu, A. Muresan, O. Constandache, G. Lisa, E.L. Muresan, C. Munteanu, et al., Tinctorial response of recycled PET fibers to chemical modifications during saponification and aminolysis reactions, Ind. Eng. Chem. Res. 53 (43) (2014) 16652–16663.
[33] A. Mittal, R.K. Soni, K. Dutt, S. Singh, Scanning electron microscopic study of hazardous waste flakes of polyethylene terephthalate (PET) by aminolysis and ammonolysis, J. Hazard. Mater. 178 (1–3) (2010) 390–396.
[34] A.M. Al-Sabagh, F.Z. Yehia, G. Eshaq, A.M. Rabie, A.E. El Metwally, Greener routes for recycling of polyethylene terephthalate, Egypt. J. Petrol. 25 (1) (2016) 53–64.
[35] C. Lorenzetti, P. Manaresi, C. Berti, G. Barbiroli, Chemical recovery of useful chemicals from polyester (PET) waste for resource conservation: a survey of state of the art, J. Polym. Environ. 14 (1) (2006) 89–101.
[36] Blackmon, K.P., Fox, D.W. and Shafer S.J., 1990. Process for converting PET scrap into diamine monomers, U.S. Patent No. 4 973 746.
[37] A. Jain, R.K. Soni, Spectroscopic investigation of end products obtained by ammonolysis of poly(ethylene terephthalate) waste in the presence of zinc acetate as a catalyst, J. Polym. Res. 14 (6) (2007) 475–481.
[38] N.D. Sharma, A.A. Vaidya, P. Sharma, Recovery of pure terephthalic acid from polyester materials. Ind. Patent 163 385; 1985. CA 112: 76613d.

7 Chemical Depolymerization of PET Bottles via Combined Chemolysis Methods

Rabindra Kumar Padhan and Anand Sreeram
Department of Civil and Environmental Engineering, Hong Kong Polytechnic University, Hong Kong, P.R. China

7.1 Introduction

Polyethylene terephthalate (PET) is widely regarded as one of the most significant semicrystalline thermoplastic polyester polymers currently in use and seen a vast surge in its application in the past few decades [1]. Due to its exceptional durability, thermal and rheological properties, it is used for a variety of purposes such as food and beverage packaging material (mainly, water bottles and soft-drink bottles), houseware, electronics, automotive parts, sports goods, and textiles [2–4]. The global consumption of PET currently aggregates to around 13 million tons and is expected to grow at a rate of 5.2% per year till 2019 [5]. However, the natural slow decay of PET materials and their poor disposal management has created a major environmental hazard when these materials are disposed after use. With regard to the ever-increasing PET material production and use, recycling plays a significant role in reducing wastes by converting into it into new chemicals, monomer, and others value-added products. The recycling of PET can be conducted using various methods such as mechanical recycling and chemical recycling [6]. Mechanical recycling of PET materials usually consists of multiple stages such as sorting, granulation, floatation, washing, drying, extrusion, and granulation of PET. However, several quality problems are associated with this method and restricts its use in the beverages and food packaging industries. Chemical recycling or chemolysis on the other hand offers the significant advantage of carrying out solvolytic chain cleavage reactions which enables to produce many value-added products. In these solvolysis reactions, PET degradation is conducted using different solvents such as water, amines, alcohol, acids, etc. to produce useful by-products. Chemolysis processes for PET are

Recycling of Polyethylene Terephthalate Bottles.
DOI: https://doi.org/10.1016/B978-0-12-811361-5.00007-9

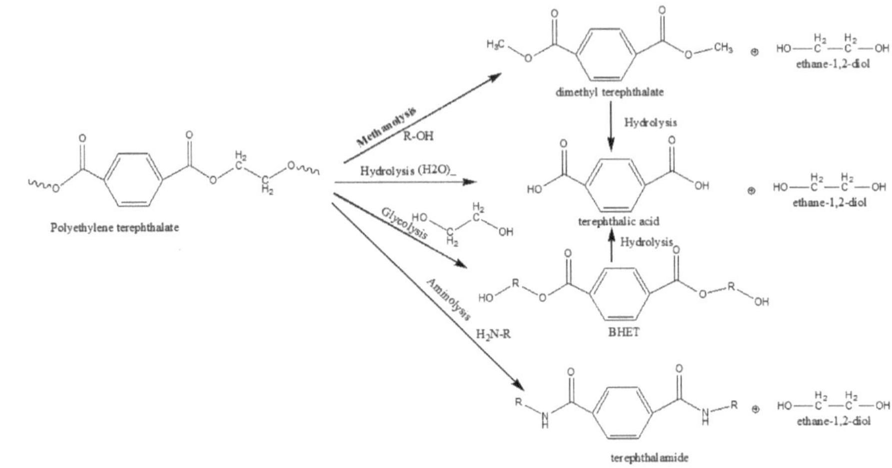

Figure 7.1 Chemistry of different chemolysis processes.

mainly divided as follows: glycolysis, hydrolysis, methanolysis, and aminolysis processes [7]. Fig. 7.1 shows the basic chemistry involved in the different chemolysis processes [8].

7.2 Various Chemolysis Processes

7.2.1 Methanolysis Process

Methanolysis of PET is generally carried out using the solvolysis reaction of PET with methanol solvent at a temperature of around 180−280°C in a high-pressure environment of 20−40 atm. The by-products of the methanolysis processes are dimethyl terephthalate (DMT) and ethylene glycol (EG) [9−12]. The main catalysts used in the methanolysis include zinc acetate, magnesium acetate, cobalt acetate, and lead dioxide. The reaction products formed during methanolysis reaction comprise an extensive combination of alcohols, glycols, and phthalate derivatives [13]. Some studies have also looked at the methanolysis of waste PET with the use of aluminum isopropoxide as a catalyst to form monomers and reported that using an amalgamation of toluene/methanol solvent (20 vol.% of toluene) in the place of methanol, can lead to better yield of monomers [14]. The main advantages of the methanolysis process are that chemicals used during the reaction process are low-cost and comparatively less hazardous to the environment [15]. However, a frequent problem that occurs is the need for separation of EG from DMT and catalyst residues, which restricts methanolysis from becoming a fully viable process.

7.2.2 Hydrolysis Process

The reaction of PET with water under various conditions breaks the polyester chains into terephthalic acid (TPA) and EG. The hydrolysis reaction is conducted using concentrated acids such as sulfuric acid, phosphoric, or nitric acid. However, concentrated sulfuric acid is the most commonly used chemical for this process [16–19]. On the other hand, alkaline hydrolysis of PET is conducted using alkaline solutions of NaOH or KOH at various concentrations [6,20,21]. The subsequent products obtained in the reaction process include EG and the corresponding terephthalate salt depending on the alkaline solution. Neutral hydrolysis of PET can also be accomplished by the means of water or steam in the presence of alkali metal acetates [22–24]. However, low purity of TPA formed after the reaction is a significant is drawback of this method.

7.2.3 Glycolysis Process

Glycolysis is the most simple and oldest method among all the PET chemolysis processes. Several well-known companies and enterprises have previously used glycolysis as their preferred PET recycling method and is hence considered to be one of the most important commercial approaches for the chemical recycling of waste PET [25,26]. In the glycolysis process, PET reacts with various glycols, such as EG and diethylene glycol (DEG). These reactions are carried out at relatively high temperature (180–240°C) which breaks the PET polymeric chains into TPA, EG, α,ω-dihydroxy materials (polyols). The α,ω-dihydroxy materials are subsequently used for the production of various polymers such as polyurethanes, polyesters, epoxy resins, and vinyl esters [27–30].

7.2.4 Aminolysis Process

Aminolysis is a method for PET chemolysis which has been relatively lesser explored in comparison to other processes. There exist only a few literature references regarding the chemolysis of the waste PET using different amines such as ethanol amine, tri-ethanol amine, allylamine, and polyamines [7,31,32]. Chemicals such as acetic acid, zinc acetate, lead acetate, and other simple salts have been found to be capable of depolymerizing PET through aminolysis reactions [31]. With regard to the aminolysis process, the yield and purity of the products are generally high and has the potential to synthesized useful products through several chemical reactions.

7.2.5 Commercial Viability

From an economic point of view, PET chemolysis via hydrolysis and methanolysis process is a very capital-intensive process, and hence it is economically not viable for small scale production. However, this is not the case for PET chemolysis via glycolysis process, which may be economically viable for small and medium scale production.

7.3 Combined Chemolysis Process

In the last decades, several new methods for PET recycling via chemical processing have been patented. These processes not only use chemical agents but also consist of two or more steps which combine distinct chemolysis methods. Some combined chemolysis processes are also highly attractive for industrial applications due of the lenient reaction conditions. Additionally, huge quantity of contaminants present in the original waste PET can be removed after the initial treatment, which favors final product decontamination from impurities. The final products produced during this process can also be further processed for polymerization reactions into PET or other materials. The most important combined chemolysis methods are: glycolysis–hydrolysis, methanolysis–hydrolysis, glycolysis–aminolysis, and glycolysis–methanolysis.

7.3.1 Glycolysis–Hydrolysis

The patent for a two-stage combined chemolysis process involving glycolysis–hydrolysis of waste PET material was granted in 1985 [33]. In this process, EG is firstly introduced into liquefied PET to decrease its molecular weight and is termed to as the preglycolysis step. When this preglycolysis step is attached with a hydrolyzer vessel, the residence time is significantly condensed. The product generated from preglycolysis step was employed to hydrolyzer vessel to complete the chemolysis reaction of PET oligomers to EG and TPA. However, PET waste processed in this manner contains a substantial amount of water, the required equilibrium residence time to insure a complete chemolysis reaction from PET to monomer requires at least 45 min. Other methods involved the reaction of PET with excess amount of EG in the presence of xylene to form oligomeric products to obtain monohydroxy ethyl terephthalate (MHT), bis(2-hydroxy ethyl)terephthalate (BHET) [34]. When the preglycolysis reaction of PET takes place, the molecular weight of PET reduces extremely fast due to the various conditions in the melt extruder. The resultant products formed in this process are not TA and EG, but rather BHET, and an assortment of other hydroxyl-terminated oligomers. The residence time in this case becomes irrelevant and no TA and EG are formed during the initial steps, i.e., preglycolysis reaction. However, the desired end products are obtained when both the processes, i.e., preglycolysis and hydrolysis are combined in series. Extruder conditions appropriate to the preglycolysis process of PET are preferably 280–310°C and at pressures 500–2000 psi. The most significant advantage of the two steps combined glycolysis–hydrolysis reaction approach is that the time needed for the two-step process is condensed from a third to a half than that for a single-step hydrolysis.

7.3.2 Glycolysis–Methanolysis

The glycolysis–methanolysis process developed comprises the initial dissolving of PET in a mixture of EG, TPA, and DMT oligomers [35]. This stream of reactants is subsequently reacted with superheated methanol at temperatures between 250°C and 290°C to form a product that that contains a high percentage of DMT. Additionally, majority of the contaminants linger in the first reactor, thus enabling the production of high purity DMT [36]. A hybrid process to recycle PET using glycolysis–methanolysis has also been developed to increase the yield of DMT and rate of production [37]. After an investigation of the various reaction parameters, it was inferred that the methanolysis step has a greater role than the glycolysis step and that the yield reaches a maximum at a temperature of 513K and mole ratio of EG to PET at 0.52.

7.3.3 Methanolysis–Hydrolysis

In the two-step methanolysis–hydrolysis process, PET is firstly reacted with superheated methanol vapors to produce DMT, monomethyl terephthalate, EG, and oligomeric products [38]. These products are then fractionally distilled and the remaining residues mainly containing oligomers are hydrolyzed in reactors at high temperatures, the TPA is then precipitated. This reaction process is complemented in around 15 min, which leads to a significant decrease in time required for the depolymerization process in contrast to the convention hydrolysis process which takes around 45 min.

Mitsubishi heavy industries LTD (MHI) developed a new commercial combined chemolysis process (methanolysis in supercritical methanol) to recycle waste PET [39,40]. Figs. 7.2 and 7.3 illustrate a brief framework and reaction model of the Mitsubishi process [39]. The process mainly consists of a shredding unit, the depolymerization unit, separation and purification unit, and lastly a hydrolysis unit. After shredding the waste PET, methanolysis is used to depolymerize it into

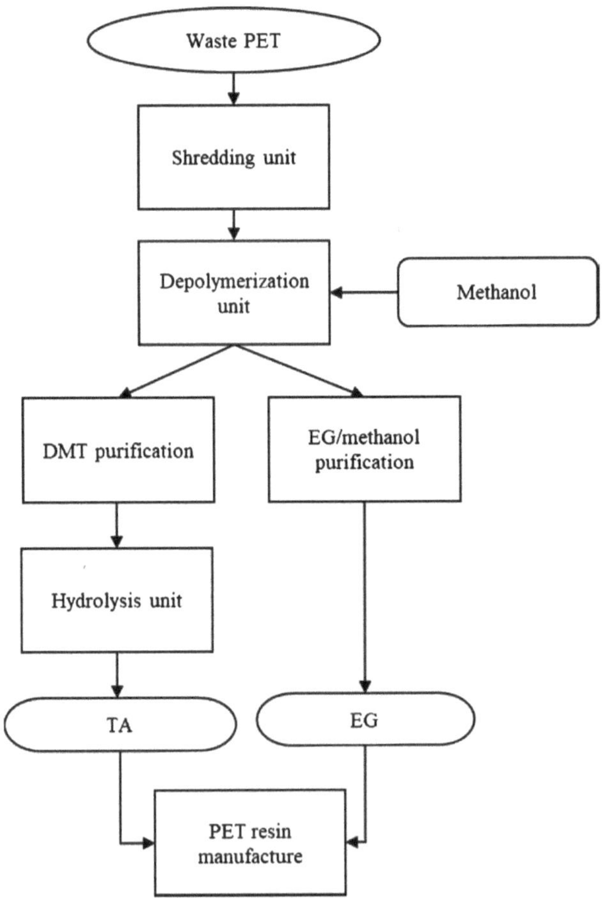

Figure 7.2 Schematic diagram of PET recycling by Mitsubishi process.

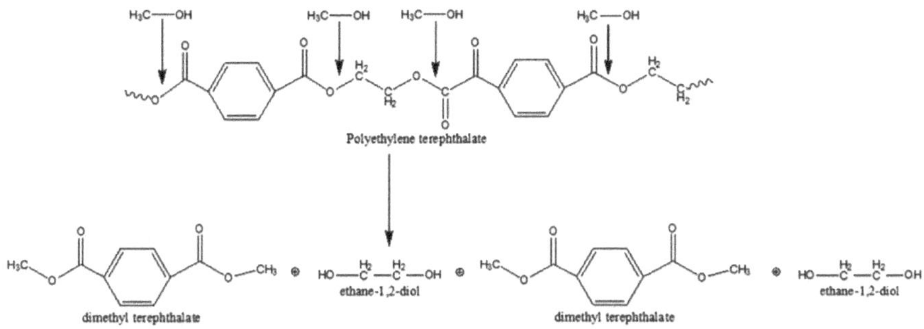

Figure 7.3 Reaction model scheme of the Mitsubishi process.

DMT and EG. Subsequently, it is distilled by fractional distillation and other chemicals. Lastly, the resultant DMT is converted into PTA in the hydrolysis unit [39–42]. The materials obtained can be then sent to production plants for further PET production. A significant advantage of such as methanolysis–hydrolysis is the efficiency of the process even when the waste PET may be contaminated by various materials.

7.3.4 Glycolysis–Aminolysis

Terephthalohydrazide based materials can be obtained from PET using a two-step chemolysis process [43]. First, waste PET in conjunction with lead acetate as a catalyst is reacted with EG, following this a product of PET with low molecular weight is obtained. In the subsequent step, cyclohexylamine, chlorobenzene, and hydrazine monohydrate are used to convert the first reaction product into terephthalohydrazide. Similar processes to formulate high yield terephthalohydroxamic acid and terephthalohydrazide have also been recently reported in literature [44]. In this, the depolymerization of PET is conducted in EG after which it is treated with hydroxylamine. The products obtained from the depolymerization reaction have been previously used for the synthesis for water-reducible alkyd resins [45,46].

7.4 Advantages and Disadvantages of Combined Chemolysis

Combined chemolysis processes offers significant advantages and has immense potential growth for future industrial application. Some of its advantages are listed as follows:

- Combined chemolysis recycling process of PET is relatively straightforward, requires less investment, and utilizes established equipment.
- Combined chemolysis such as glycolysis–hydrolysis process is advantageous over conventional methods due of the lenient chemolysis conditions in the initial step.
- Combined chemolysis methods such as glycolysis–hydrolysis favor final product purity.
- Methanolysis–hydrolysis processes are more time efficient in comparison to conventional hydrolysis.

Despite several advantages of the combined chemolysis process, it still poses several barriers including:

- Trace metals such as antimony, cobalt, and manganese present in the waste PET may lead to side reactions and unwanted products.
- Fillers and additives existing in waste PET might have an undesirable effect on the overall process, making it challenging to use the solvent for a lengthier period without regeneration.
- Combined chemolysis of PET has not been widely reached as compared to other conventional chemolysis process. There is a need of further research in order to establish the process and optimize reaction conditions.

7.5 Summary of Products From Chemolysis

Figs. 7.4 and 7.5 show the different processes for PET chemolysis and combined chemolysis including the various products formed [47]. The main products obtained from the methanolysis process include DMT and EG. The DMT produced can be further purified to synthesize new PET [48]. The aminolysis of PET leads to the formation of various terephthalamide derivatives [49,50]. In the glycolysis reaction, PET can be depolymerized to obtain various oligomeric diols and polyols, or glycolyzed into its monomeric units, BHET or DMT [51]. Combined glycolysis–hydrolysis can be used to depolymerize PET into TPA and EG, BHET, and variety of other hydroxyl-terminated oligomers. Combined methanolysis–hydrolysis reactions of PET produce DMT, monomethyl terephthalate, EG, and oligomeric products. Lastly, combined

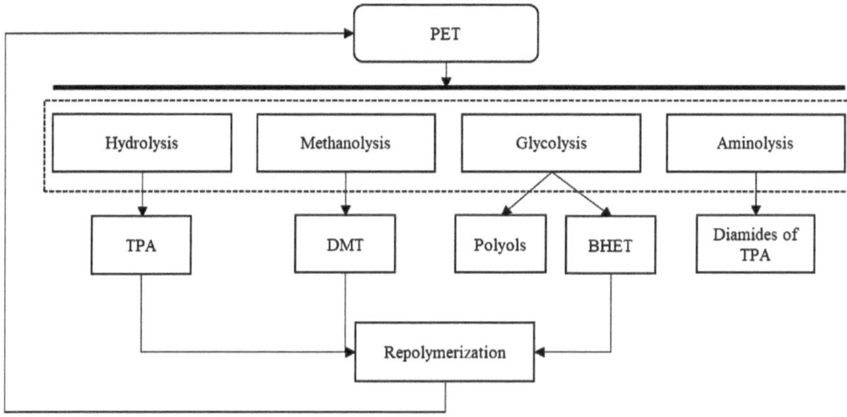

Figure 7.4 Types of PET chemolysis and product formed.

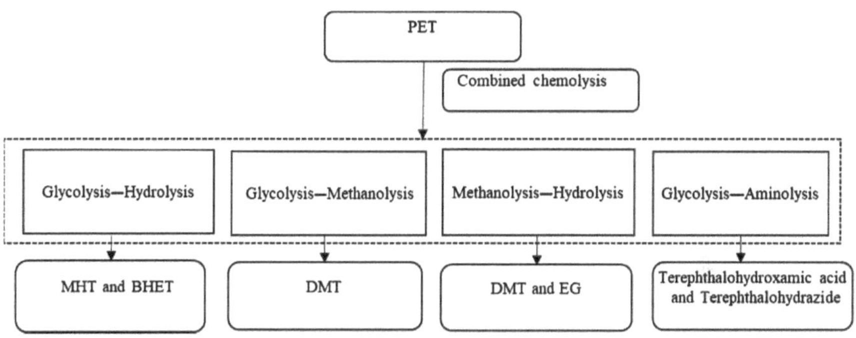

Figure 7.5 Types of combined chemolysis process of PET and product formed.

glycolysis—methanolysis process also produces high purity DMT. Oligoesters from the glycolysis—hydrolysis of waste PET can be further used in the manufacture of various polymers such as polyurethanes and plasticizers. The other products obtained can be used for applications such as UV curable coatings, water-reducible alkyd resins, and epoxy-based paints [52].

7.6 Conclusion

This chapter provides a brief summary regarding the various chemolysis methods for the recycling of waste PET including conventional methods such as: hydrolysis, methanolysis, glycolysis and aminolysis and two-step methods which are a combination of the one-step methods in series. Overall, combined chemolysis processes offer a relatively simple, efficient and effective method as compared to traditional methods. There also exists immense potential for future industrial and commercial applications using such chemolysis methods. However, extensive research is still required to cement this technology as profitable and economically viable. It should be noted that although a variety of chemolysis methods stated in this chapter are available for recycling PET, in reality only a small fraction of waste PET is currently being processed through these methods. Therefore, there subsists a significant need for researchers and professionals to widely commercialize these methods, and help curtail this waste disposal problem.

References

[1] S.V. Levchik, E.D. Weil, A review on thermal decomposition and combustion of thermoplastic polyesters, Polym. Adv. Technol. 15 (2004) 691–700.
[2] F. Awaja, D. Pavel, Recycling of PET, Eur. Polym. J. 41 (2005) 1453–1477.
[3] D. Kint, S. Muñoz-Guerra, A review on the potential biodegradability of poly(ethylene terephthalate), Polym. Int. 48 (1999) 346–352.
[4] R.J. Caldicott, The basics of stretch blow molding PET containers, Plast. Eng. 35 (1999) 35–38.
[5] Polyethylene Terephthalate (PET): 2015 World Market Outlook and Forecast up to 2019.
[6] V. Sinha, M.R. Patel, J.V. Patel, PET waste management by chemical recycling: a review, J. Polym. Environ. 18 (2010) 8–25.
[7] D. Paszun, T. Spychaj, Chemical recycling of poly(ethylene terephthalate), Ind. Eng. Chem. Res. 36 (4) (1997) 1373–1383.
[8] C. Lorenzetti, P. Manaresi, C. Berti, G. Barbiroli, Chemical recovery of useful chemicals from polyester (PET) waste for resource conservation: a survey of state of the art, J. Polym. Environ. 14 (1) (2006) 81–101.
[9] Lotz, R., Wick, G., and Neuhaus, C., 1967. U.S. Patent 3 321 510.
[10] Gruschke, H., Hammerschick, W., and Medem, W., 1968. U.S. Patent 3 403 115.
[11] M. Goto, H. Koyamoto, A. Kodama, T. Hirose, S. Nagaoka, Depolymerization of polyethylene terephthalate in supercritical methanol, J. Phys.: Condens. Matter. 14 (2002) 11427–11430.
[12] M. Genta, T. Iwaya, M. Sasaki, M. Goto, T. Hirose, Depolymerization mechanism of poly(ethylene terephthalate) in supercritical methanol, Ind. Eng. Chem. Res. 44 (2005) 3894–3900.

[13] Cornell, D.D., 1993. Depolymerization of PET for food packaging, in: Proceedings of the Conference on Recycling 'RETEC', Society of Plastic Engineers, Bethel, CT, USA.
[14] H. Kurokawa, M. Ohshima, K. Sugiyama, H. Miura, Methanolysis of polyethylene terephthalate (PET) in the presence of aluminum tiisopropoxide catalyst to form dimethyl terephthalate and ethylene glycol, Polym. Degrad. Stab. 79 (2003) 529–533.
[15] S.R. Shukla, A.M. Harad, L.S. Jawale, Recycling of waste PET into useful textile auxiliaries, Waste Manag. 28 (1) (2008) 51–56.
[16] Pusztaszeri, S.F., 1982. U.S. Patent 4 355 175.
[17] G.P. Karayannidis, A.P. Chatziavgoustis, D.S. Achilias, Poly(ethylene terephthalate) recycling and recovery of pure terephthalic acid by alkaline hydrolysis, Adv. Polym. Technol. 21 (2002) 250–259.
[18] C.G.G. Namboori, M.S. Haith, Steric effects in the basic hydrolysis of poly(ethylene terephthalate), J. Appl. Polym. Sci. 12 (1968) 1999–2005.
[19] Brown, G.E. Jr., O'Brien, R.C., 1976. U.S. Patent 3 952 053.
[20] D. Paszun, T. Spychaj, Chemical recycling of poly(ethylene terephthalate), Ind. Eng. Chem. Res. 36 (1997) 1373.
[21] J. Scheirs, Recycling of PET, Polymer Recycling: Science, Technology and Applications. Wiley Series in Polymer Science, Wiley, Chichester, UK, 1998.
[22] Mandoki, J.W., 1986. U.S. Patent 4 605 762.
[23] Rosen, B.I., 1992. U.S. Patent 5 095 145.
[24] Royall, D.J., and Harvie, J.L., 1993. Eur. Patent 550 979.
[25] Simonaitis, T., Barkauskas, R., Jankauskaitė, V., 2005. Adhesive composition with poly(ethylene terephthalate) waste, in: Proceedings of the 4th Central European Conference on Plastics Recycling Science-Industry, Nitra, VUSAPL, pp. 1 – 3.
[26] J. Scheirs, Polymer Recycling: Science, Technology and Applications, Wiley, Chichester, UK, 1998.
[27] M.Y. Abdelaal, T.R. Sobahi, M.S.I. Makki, Chemical transformation of PET waste through glycolysis, Constr. Build. Mater. 25 (2011) 3267–3271.
[28] D.E. Nikles, M.S. Farahat, New motivation for the depolymerization products derived from poly(ethylene terephthalate) (PET) waste: a review, J. Macromol. Mater. Eng. 290 (2005) 13–30.
[29] S.D. Desai, J.V. Patel, M.R. Patel, V.K. Sinha, PET waste based polyurethane adhesive for rubber joints, Indian J. Chem. Technol. 12 (2005) 82–87.
[30] T. Spychaj, Handbook of Thermoplastic Polyesters: Homopolymers, Copolymers, Blends, and Composites, Wiley-VCH, Weinheim, Germany, 2005.
[31] S.R. Shukla, A.M. Harad, Aminolysis of polyethylene terephthalate waste, Polym. Degrad. Stab. 91 (2006) 1850–1854.

[32] R.K. Padhan, A.A. Gupta, R.P. Badoni, A.K. Bhatnagar, Poly(ethylene terephthalate) waste derived chemicals as an antistripping additive for bitumen—an environment friendly approach for disposal of environmentally hazardous material, Polym. Degrad. Stab. 98 (12) (2013) 2592–2601.

[33] M.L. Doerr, 1986. U.S. Patent 4 620 032.

[34] G. Güçlü, T. Yalçınyuva, S. Özgümüş, M. Orbay, Simultaneous glycolysis and hydrolysis of polyethylene terephthalate and characterization of products by differential scanning calorimetry, Polymer 44 (25) (2003) 7609–7616.

[35] Gamble, W.J., Naujokas, A.A., and DeBruin, B.R., 1994. U.S. Patent 5 298 530.

[36] M.E. Rogers, T.E. Long (Eds.), Synthetic Methods in Step-Growth Polymers, John Wiley & Sons, Inc., Hoboken, New Jersey, 2003.

[37] Bo-kyung Kim, D. Kim, Y. Cho, M. Han, Chemical recycling of poly (ethylene terephthalate) using a new hybrid process, J. Chem. Eng. Jpn. 41 (9) (2008) 923–928.

[38] J. Aguado, D. Serrano, Feedstock Recycling of Plastic Wastes, Royal Society of Chemistry, Cambridge, 1999, p. 34.

[39] Mitsubishi Heavy Industries, Ltd., Technical Review, 40 (Extra No.1), January 2003.

[40] A.S. Goje, S.A. Thakur, T.M. Patil, S. Mishra, Glycolytic aminolysis of poly (ethylene terephthalate) waste for recovery of value-added comonomer at atmospheric pressure, J. Appl. Polym. Sci. 90 (12) (2003) 3437–3444.

[41] A. Schoengen, JP Patent S57-53332.

[42] Genta, M., Uehara, R., Yano, F., and Kondo, Y., Development of Chemical Recycling Process for Post-Consumer PET Bottles by Methanolysis in Supercritical Methanol, Mitsubishi Heavy Industries, Ltd. Kobe Shipyard & Machinery Machinery & Environment Plant Department (www.isasf.net/fileadmin/files/Docs/Versailles/Papers/Md7.pdf).

[43] Y. Makoto, H. Toshiki, Y. Koji, K. Yoshio, C. Namiko, I. Tomoyuki, et al., Chemical recycling of poly(ethylene terephthalate). 2. Preparation of terephthalohydroxamic acid and terephthalohydrazide, Ind. Eng. Chem. Res. 41 (16) (2002) 3993–3998.

[44] N. George, T. Kurian, Recent developments in the chemical recycling of postconsumer poly(ethylene terephthalate), Ind. Eng. Chem. Res. 53 (2014) 14185–14198.

[45] I. Acar, A. Bal, G. Guclu, The use of intermediates obtained from aminoglycolysis of waste poly(ethylene terephthalate) (PET) for the synthesis of water-reducible alkyd resin, J. Chem. 91 (2013) 357–363.

[46] E. Bulak, I. Acar, The use of aminolysis, amino-glycolysis, and simultaneous aminolysis–hydrolysis products of waste PET for production of paint binder, Polym. Eng. Sci. 54 (2014) 2272. Available from: https://doi.org/10.1002/pen.23773.

[47] E. Rusen, A. Mocanu, F. Rizea, A. Diacon, I. Calinescu, L. Mititeanu, et al., Post-consumer PET bottles recycling II. PET depolymerization using microwaves, Mater. Plast. 50 (3) (2013) 201–207.
[48] C.H. Chen, C.Y. Chen, Y.W. Lo, C.F. Mao, W.T. Liao, Studies of glycolysis of poly (ethylene terephthalate) recycled from postconsumer soft-drink bottles. I. Influences of glycolysis conditions, J. Appl. Polym. Sci. 80 (2001) 943–948.
[49] R.K. Padhan, A.A. Gupta, Preparation and evaluation of waste PET derived polyurethane polymer modified bitumen through in situ polymerization reaction, Constr. Build. Mater. 158 (2018) 337–345.
[50] Z. Leng, R.K. Padhan, A. Sreeram, Production of a sustainable paving material through chemical recycling of waste PET into crumb rubber modified asphalt, J. Clean. Prod. 180 (2018) 682–688.
[51] G. Guclu, Alkyd resins based on waste PET for water-reducible, Polym. Bull. 64 (2010) 739–748.
[52] V.A. Kosmidis, D.S. Achilias, G.P. Karayannidis, Poly (ethylene terephthalate) recycling and recovery of pure terephthalic acid. Kinetics of a phase transfer catalyzed alkaline hydrolysis, Macromol. Mater. Eng. 286 (2001) 640–647.

8 Life Cycle Assessment (LCA) of PET Bottles

K.V. Marathe, Karan R. Chavan and Pranav Nakhate
Department of Chemical Engineering, Institute of Chemical Technology, Mumbai, India

8.1 Goal Definition Scope

8.1.1 Background

Life cycle analysis (LCA) is a way of quantifying the environmental impact generated by a product or a process [1]. Polyethylene terephthalate (PET) is a product used worldwide and mainly used for the manufacture of synthetic fibers. Around 30% of the total production is used in the packaging of foods and beverages [2]. PET bottles are commonly used for the packaging of bottled water, soft drinks, pharmaceuticals, and cosmetics. Due to its unique advantages of being an inert barrier for external gas or air and other liquids, it is primarily used in the packaging of liquid products for human consumption. LCA of such a product provides the environmental impact associated with the production, consumption, and recycling of a PET bottle. LCA is mainly conducted for informed decision-making based on the results and its interpretation. A lot of insight regarding the process and product is gathered from such a study and helps in future optimization of various applications of the same. As the world trend now is to reduce various footprints related to consumption, such analysis are carried out worldwide as they provide current and critical information related to the product and process. The aim of such study is to assist the decision makers within an organization and the concerned policy makers, executers. It also contributes to the general awareness of the public regarding the packaging industry and its contribution to the environmental impact. The suggestions and insights generated through such studies direct the future research toward the sustainability of the packaging industry [3,4].

Recycling of Polyethylene Terephthalate Bottles.
DOI: https://doi.org/10.1016/B978-0-12-811361-5.00008-0

8.1.2 Introduction

Enormous amounts of food, drinks, and medicines are packaged, stored, and transported for human consumption. Primary purpose is to deliver an unaffected, undamaged product to the end consumers. Through years of innovative development in the packaging field and trial and error we have zeroed down on a few options that will not hinder with the condition of the items being packaged. Mostly inert materials such as PET, aluminum, glass, and paper (tetra packs and cartons) are a few such options that are used all over the globe. These materials are used as per requirements of the material to be packaged and primarily depend on the economic parameter in question. Different researchers, agencies, and consultants have calculated the environmental impact of these packaging options by using various methodologies and empirical formulas. Life cycle assessment (LCA) is a tool used for assessing the environmental impact by various products and processes. Assessing the impact and knowing the value of environmental burdens being produced by packaging is the question of the hour. Such an insight will help in deciding the nature of the packaging industry that is currently a humongous industry. The quantum of PET being produced to cater to the need of packaging of medicines, water, soft drinks, cosmetics, and other food material is growing daily compared to any of the other packaging options, and hence a study of this nature will help in providing a sustainable solution.

Importance of PET as a primary packaging material in the pharmaceutical, food, and beverage industry is due to certain advantages and inherent properties of the material such as:

1. It can be recycled completely.
2. It is lightweight, elastic, insulating, and durable [5].
3. It is a low-cost product [6].
4. It is inert in nature and does not interact with the constituents/chemicals present in the food and beverages, etc. [7].
5. It is produced at a lower temperature than glass or metal packaging.
6. It has a good shelf-life and does not degrade.
7. It is transparent material hence contents can be seen by the consumer which increases the trust in the product [7].
8. It is an excellent barrier to other gases and liquids.
9. It is resistant to microbial and biochemical attack [8].
10. The PET polymer molecule chain is capable of being stretched from one or both the directions and hence

making use of this property and by different manipulations, various applications can be found for opaque PET, crystalline PET, etc. [9,10].

8.1.3 Purpose

Majority of the LCA studies are conducted to find the environmental impact associated with a certain quantity of PET production associated with specific packaging application, for example, soft drinks, alcoholic drinks, cosmetics, pharmaceutical, and medicines, and also to assess the environmental impact associated with different stages of the PET bottle life cycle such as production, transportation, and recycling. It is conducted for informed decision-making regarding a specific application or pertaining to a particular stage of the life cycle. Such study helps in chalking out policies and protocols to be followed to reduce or curb environmental burdens being produced due to myopic planning. These studies are conducted to promote sustainable decision-making regarding the disposal of waste generated by PET post consumption. The various options of landfill, incineration, or recycling are compared and the least burdensome option is promoted for better sustainability [11–13]. The option of open loop and closed looped recycling of PET bottles may drive the recycling industry in the direction of sustainability [14,15]. The further use of such studies is to compare different production processes or comparison of different unit operations in order to find the least impactful thus contributing in reducing the footprint of the PET production [15,16]. By applying the policy of "prevention comes before recycling comes before waste disposal" the application of LCA tool can be applied at various levels and stages in order to get clarity and achieve sustainability [17]. Various recycling systems, protocols, and policies can be designed from the insight gained at each stage, post consumption of the PET bottles. An incentive-based recycling system is already in place in countries such as Sweden and Switzerland which in turn helps in better segregation of the waste bottles and optimized recycling can be carried out in context of textile production, etc. [17].

8.1.4 Previous Research

Expert consultants and researchers pertaining to specific geographical regions, countries, or organizations have conducted various LCA studies. The same has been carried out to assess the region-specific impact. Other studies are conducted to know the impact being produced by various stages of the life cycle of the PET bottle—cradle to grave. Such studies provide a product life cycle insight which in turn will help in achieving sustainability.

Different experts in sustainability such as PE International, Franklin Associates, and universities such as Michigan, Stuttgart, and Utrecht have conducted various studies in this domain pertaining to LCA of plastics packaging for the United States and Canada, LCA of polymers in automotive industry, comparing polylactic acid (PLA) and PET bottles, LCA of containers for olive oil, LCA of different unit operations such as injection molding and thermoforming, production of high-density polyethylene (HDPE) and PET resin postconsumers, comparison of nine plastic resins, comparison of PET with aluminum and paper containers, LCA of tap versus bottled water, LCA of wine containers, i.e., glass versus PET, etc.

Also LCA for PET bottles has been carried out for different countries and regions such as Canada, the United States, Oregon, California, Brazil, the Netherlands, Germany, Mexico, India, and Mauritius, thus providing a better judgment of the environmental impact due to the packing industry. In 1969, the Midwest Research Institute first carried out LCA of packaging options for Coca-Cola and later others followed the suit to reach conclusions regarding the same problem in different regional context [18].

8.1.5 Market Trends

Climate change and environmental hazards are the most important questions posed in front of mankind. As a result of exponential growth of human population the total consumption of primary requirements has risen along with in the past decade. Bottled water will be the most packed product entity by 2019 according to CAGR forecast and will surpass cigarettes, drinking milk products, and baked products. As PET has proved itself of being the most versatile packaging material, its total production in 2015–16 was ~1450 kT of PET in India as compared to ~980 kT in 2014–15 (against a capacity of 1326 kT). There are four major manufacturers of PET in India: Reliance Industries Limited, Dhunseri Petrochem and Tea Limited, JBF Industries Limited, Micro Poly Pet Private Limited (now Indorama), and the total installed capacity of production installed is 1976 kT. This has risen from a 1326 kT in 2014–15. A large quantity of PET is exported every year. In 2015–16, ~650 kT of PET was exported. The major markets for PET exports from India are Bangladesh, the United States, Italy, Israel, Romania, Ukraine, UAE, etc. The export volumes have grown in the recent years, closely tracking the overall production levels in India. Maximum use of a PET bottle is for packaging of water followed by carbonated soft drinks, tea, juice, energy drinks, cosmetics, food, and other home care products. The total consumption of PET bottles was 488 billion units in 2016 and an estimated 87.5 billion units of growth is forecasted for bottled water till 2019 mainly in the Asia Pacific region.

8.1.6 Need for the Project

Climate change and environmental hazards are the most important questions being posed in front of mankind. Most of the environmental burdens have been caused by unsustainable growth. Products and processes have been designed keeping in mind only the commercial aspect. By-products and emissions of a process have been ignored for not having any commercial benefit in disposing or treating the same. A truly sustainable process will neither have any by-products nor any emissions. The natural resources available have been exploited to incur commercial benefits by a select few. Lack of holistic understanding regarding the life cycle of process or product and dearth of interdisciplinary communication between policy makers and process/product designers have resulted in many inefficiencies translating into environmental impact.

8.1.7 Targeted Audience and Use of the Study Product System

Policy makers and decision heads of an organization and government are generally the targeted audience for such studies. As the general population is becoming aware about environmental hazards being caused by various day-to-day products, thus such studies foster the use of environmental-friendly products. As people from all walks of life are affected by the packaging industry and most of the products are economically driven the environmental aspect in the product selection would be a decisive step in the current era. New product designers would be driven by the insights obtained from such studies. Designing a protocol or a system to eradicate the nuisance caused by a product at a stage in its life cycle can very well be based on LCA studies. The key findings in such a study enable in fixing protocols based on a macro level understanding.

8.1.8 Functional Unit

Functional unit is the basis to conduct the LCA study for a particular product or process. It is the central entity around which the study revolves making it the most important aspect of the assessment. In majority of cases the same is normally a particular sized bottle in case of PET allied studies. The life cycle inventory data is collected in context of the functional unit of the study. The primary criteria being the environmental impacts should be related to the functional unit chosen. Mostly as the purpose differs for each study based on the motive of the study being conducted by different organizations and experts the functional unit varies. As we can see in Tables 8.1 and 8.2 most of the studies consider a size of PET bottle as the functional unit. Other studies have also considered their total usage or production of the product as the functional unit for the LCA study. During comparative studies, the entities under consideration are commonly the units used as single serving of different materials of packaging.

Table 8.1 Review of Different LCA Studies

S. No.	Title of the Study	Agencies
1	LCI Summary for PLA and PET 12-Ounce Water Bottles	Franklin Associates, 2007 [18]
2	Life Cycle Assessment of PET Bottle	Dokuz Eylül University, 2008 [6]
3	Life Cycle Assessment (LCA) of Pet Bottles and Comparative LCA of Three Disposal Options In Mauritius	University of Mauritius, 2008 [19]
4	Life Cycle Inventory of Three Single-Serving Soft Drink Containers	Franklin Associates, 2009 [20]
5	Life Cycle Assessment of Drinking Water Systems: Bottle Water, Tap Water, and Home/Office Delivery Water	Franklin Associates, 2009 [10]
6	Life Cycle Assessment of Example Packaging Systems for Milk	Environmental Resources Management Limited, 2010 [21]
7	Life-Cycle Assessment Summary Report	Corrugated Packaging Alliance, February 2010 [22]
8	Life Cycle Inventory of 100% Postconsumer HDPE and PET Recycled Resin from Postconsumer Containers and Packaging	Franklin Associates, A Division of Erg Prairie Village, Kansas, 2011 [5]

Table 8.1 Review of Different LCA Studies—cont'd

S. No.	Title of the Study	Agencies
9	Life Cycle Assessment of Polyethylene Terephthalate (PET) Beverage Bottles Consumed in the State of California	California Department of Resources Recycling and Recovery, 2011 [23]
10	Comparative Life Cycle Assessment Report of Food Packaging Products	Cascades Specialty Product Groups, Quebec, Canada, 2011 [24]
11	Life Cycle Assessment of Polymers in an Automotive Assist Step	PE International, Inc., 2012
12	An Analysis of Life Cycle Assessment in Packaging For Food & Beverage Applications	United Nations Environment Programme and Setac, 2013 [4]
13	Life Cycle Assessment of Pet Bottle Recycling: A Case Study for Mexico	Vienna School of International Studies, 2013 [7]
14	Glass or Plastic: An Environmental Life Cycle Assessment (LCA) and Related Economic Impact of Contrast Media Packaging	European Society of Radiology, 2015 [25]
15	Environmental Evaluation of Non-Alcoholic Single-Serve PET Beverage Bottles in the State of California Using Life Cycle Assessment and System Dynamics	Michigan State University, 2015 [26]
16	LCA of Selected Packaging Products	Ecopaperloop Project by European Union (EU) [27]

8.1.9 Assessment Boundaries

As previously this was a most debated part of a life cycle assessment study, certain protocols and standards are recommended by the ISO 14040: 2006 guidelines [39,40]. Majority of the studies clearly define the goal and scope of the study, thus making it easy for the reviewers to understand the various criteria considered in each one of them. Since various assumptions and constraints define the scope of the LCA study

Table 8.2 Functional Units Considered in Different LCA Studies

S. No	Functional Unit	Reference
1	1 L beverage bottle	[28,12,17]
2	1 kg PET resin	[29,30,23]
3	350 kg of PET bottles and 650 kg of PET fibers	[31]
4	100 bottles of 0.5 L	[32]
5	Two 50 mL, one 100 mL, half of a 200 mL and one-fifth of a 500 mL bottles	[25]
6	6000 bottles of 1.5 L capacity	[19]
7	12 ounce PET bottle	[10,20,18]
8	1 m^3 of water in bottles	[33]
9	1000 units of 500 mL bottles	[34]
10	1 metric tonne of PET fibers	[35]
11	1000 L carbonated soft drink beverage	[26]
12	Production, collect and treatment of PET bottles for distributing 1000 L of beverage to consumer	[36]
13	1000 units of 2 L bottle	[7]
14	1 L of polymer contained in pellet (prior to product molding)	[37]
15	60 kg PET bottles	[38]
16	1000 units of 1.5 L PET bottle	[13]

the same are chosen pertaining to the first-degree association or linkage with the life cycle of the product or process. The system boundary is clearly mentioned and the further scenarios are explored based on various possibilities that can take place given the application of the product in real life. The same can be investigated through sensitivity analysis by taking various scenarios into consideration. Also these guidelines are not all encompassing and have their shortcomings where concerns regarding neglecting of the small product streams are raised. The LCA professional needs to have a holistic understanding of the LCA practice as well as the product under consideration in order to decide the boundary and the possible effect on the impact values. Regional or geographical boundaries can also be one of the criteria for boundary selection. Based on the scope of the study either the complete life cycle is considered or a part of the life cycle of a specific stage referred to as "cradle to gate" (raw material procurement), "gate to gate" (production stage), or "gate to grave" (product disposal) can be carried out. The input–output flow mapping for of a product or the functional unit will give a comprehensive understanding regarding the entities to be considered in the study.

8.1.10 General Exclusions

The system boundary for the LCA study is defined during the initial stages based on the goal and scope. General exclusions in the study are the impacts being contributed by the building construction of the production facility, production of the process vessels and other peripheral machinery required in the treatment of the raw materials. The environmental impact being produced by the number of people involved during the production process. The impact being produced by the production of vehicles used for transportation of raw materials and finished products. The comprehensiveness of the study depends on the boundary and scope of the study. The abovementioned impacts are not considered in majority of the studies as it increases the complexity of the study.

In order to conduct a comprehensive life cycle assessment of a product or process, various factors and all the dependent processes throughout the life cycle have to be considered. Most of the LCA studies define their boundary conditions to reduce the complexity of the study and consider the immediate associated processes. For example, PET bottle production considers raw materials required and obviates from considering the impact contributed by the production of pressure vessels and machinery or the steel used to produce them required to do so. As the chain is interlinked, LCA of all the materials will have to be considered to find the impact of a PET bottle. Majority of the studies define the boundary conditions till the first-degree linkage with the material under consideration. Most of the studies do not consider the mass of the flow of materials which amount to be $<1\%$ of the input–output inventory. The environmental impact of such products needs to be irrelevant in order to ignore the flow based on mass criteria. Most of the catalysts and additives fall under this category during the PET production.

8.2 Life Cycle Inventory

8.2.1 General Methodology

The total life cycle of a product or process is divided in the stages for the ease and better understanding of the study. The same also aids the LCA professional in deciding the action plan for the execution of the study. Defining goal and scope that will decide the system boundary for the study is the first step. The next step followed is inventory data collection and its analysis. It can be seen that the input–output flow mapping for a certain product will generate a format for data collection. The consolidated inventory data for various stages of life cycle of a product is checked for consistency and if any gaps found the same is either revised or particular assumptions are made regarding the missing data. Life cycle inventory datasets are checked and appropriated for all the material and energy flows involved in the study of the life cycle. These datasets are provided by agencies such as NREL, Ecoinvent, etc. and most of the LCA software providers such as GaBi, Simapro, Umberto, etc. have their own datasets or other datasets can be imported in the same for different materials and energy flows. In the impact assessment step of LCA the environmental impact categories to be analyzed are chosen based on the requirement of the study. Various impact assessment methods are present such as CML2001, TRACI 2002, Eco-indicator, etc. that can be used in a standalone or combination. The standard midpoint and end-point assessment methods have impact categories showcasing the effect across major sections namely resource (renewable, non-renewable, abiotic, etc.), human (global warming potential 20,100; human toxicity, etc.), land (terrestrial ecotoxicity, land use, hazardous and non-hazardous landfill waste, etc.), air (ozone depletion, photochemical oxidation, etc.) and water (marine sediment ecotoxicity, fresh water sediment ecotoxicity). Normalization is carried out of the existing impact categories by using a factor published in literature and is chosen depending on the context of the LCA study. Normally it is carried out to achieve a dimensionless number thus giving an impact value across all abovementioned sections. Further, the interpretation of the results obtained across the impact categories considered is performed based on which sensitivity and uncertainty analysis is carried out. The LCA methodology is defined in the ISO 14040: 2006 and 14044: 2006 normally to regulate the various LCA studies carried out in the world [12,37]. This standardization is crucial in order to provide a common ground to all the analysis carried out in this domain.

8.2.2 Upstream and Downstream Life Cycle Methodology

In the life cycle of a PET bottle the stages leading to its production is termed as the upstream life cycle whereas the stages pertaining to its collection and disposal are regarded as downstream life cycle. The LCA of PET bottle can be divided in various stages for ease of understanding and data collection. Such bifurcation ends up simplifying the complex life cycle of the PET bottle. The various stages are termed as "Gates" and thus provide a structure to the study. The first stage being Cradle to Gate that comprises of raw material acquisition where the resources needed for the production of raw materials required for the production PET is taken into consideration. The raw materials are transported to the factories where they are mixed and heated and are subjected to unit operations such as melt polymerization and solid state polymerization at high temperatures. The resin beads are molded in the plastic fabrication stage in preforms according to the various sizes of PET bottles. The unit operation involves high temperature and specialized machinery. The same preforms are then transported to the sites where they are blow molded in full sized bottles that are further filled with the constituents of the same depending on the application. For packaging of pharmaceutical products, clean room setups are used to avoid any microbial contamination. This Gate to Gate stage of the life cycle where the production process of PET takes place often is a major contributor to the impact footprint. Waste collection, recycle, and disposal are considered the Gate to Grave Stage, postconsumption of PET bottles. As seen in Fig. 8.1, the life cycle of the PET bottle can be seen as a closed loop when the waste is recycled back to production of bottles. The same can be sent for the production polyester fiber fill, spinning fiber, partially oriented yarn, sheets, and straps, etc.

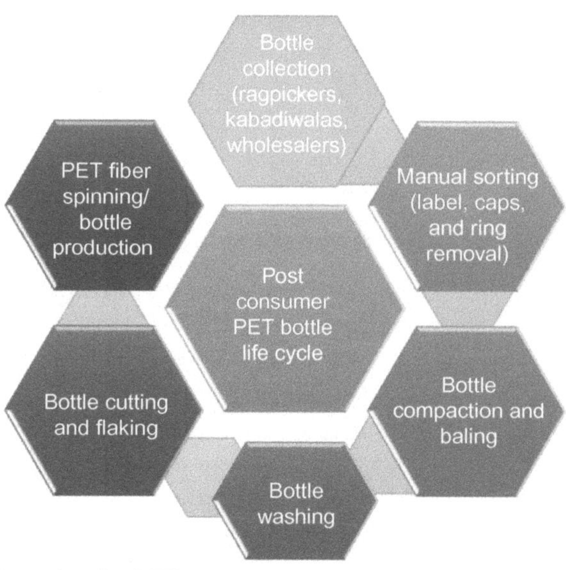

Figure 8.1 Life cycle of a PET bottle.

8.2.3 Manufacturing Stage

The raw materials required for the production of PET are shipped to the production facility. Pet production involves either an esterification or a transesterification reaction. The initial reactants in each case are either ethylene glycol and terephthalic acid or ethylene glycol and dimethyl terephthalate [2,7,19]. The former gives mere water as a by-product whereas latter generates methanol that has to be removed via distillation to take the reaction forward. The reactions are carried out in autoclaves at temperatures of 260–2700°C and 2000°C, respectively. Excess ethylene glycol is removed with the help of vacuum distillation. Bis-hydroxyethyl terephthalate is produced as the intermediate that further is polymerized and exists in the liquid form giving amorphous PET [2]. Based on the unit operations a detailed process data input format is circulated with the process engineers. The format mainly consists of the quantum of material flow, the utilities being used for fulfilling the energy demand. Input data also consists of the process parameters pertaining to different unit operations involved in the production process. The output data comprises of the product quantum and other by products being generated, the effluents from the production process [2]. The emissions from the production process are also mapped in this exercise. Auxiliary/ancillary data consists of the manpower involved in the activity and the time of operation. Consistent unit system (e.g., SI) is maintained throughout for the various material and energy flows in the system. After the procurement of the data, it is analyzed for consistency and if required more data is demanded depending on the gaps found in the inventory data.

Process datasheet for the input–output flow is prepared according to the process flow diagram shown in Fig. 8.2.

8.2.4 Product Usage and Recycling Stage

PET bottle recycle comprise of various stages postconsumption as shown in Fig. 8.3. Since the application of PET bottles is varied a large group of end users are involved in the consumption and depending on the purpose the number of usages of the bottle is decided. After the usage most of the bottles are thrown in the dust bins which are then collected by the central authority in charge of waste collection. Most of the regions in the world have a waste sorting and collection system in place. In context of India a chain of collectors operating at various levels is involved making the activity comprehensive in nature.

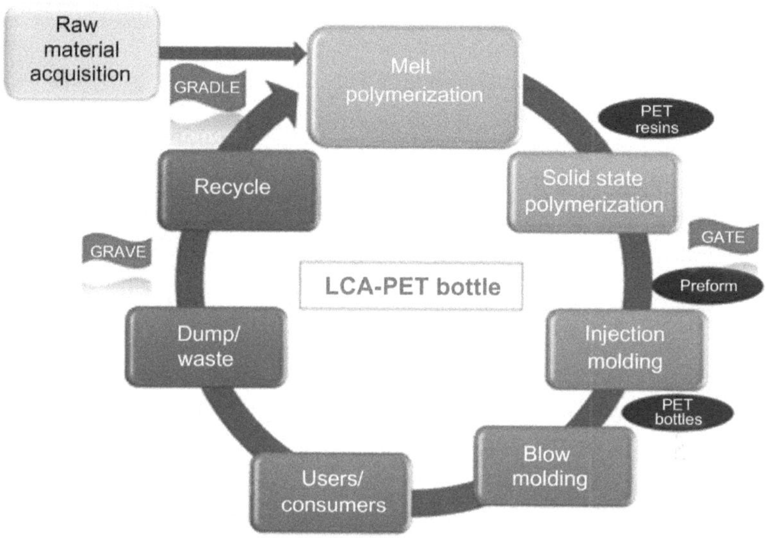

Figure 8.2 Process flow diagram for PET bottle production.

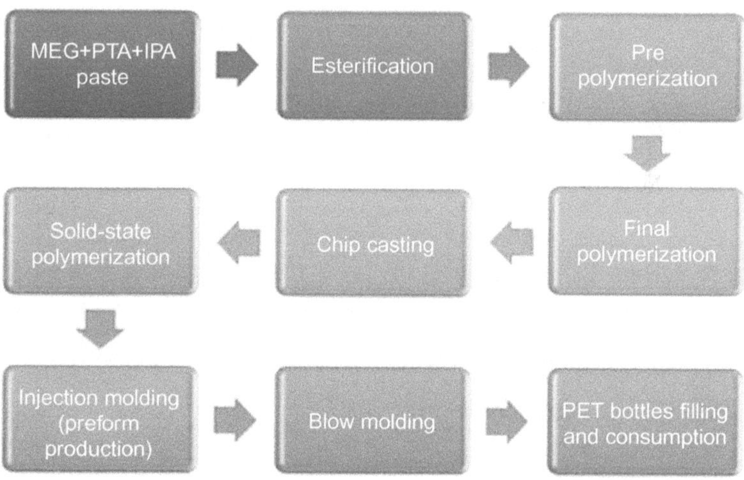

Figure 8.3 Downstream life cycle of PET bottle postconsumption.

Waste collectors being the back bone of the activity collect the PET bottles (sold for 14–15 INR/kg) which are then consolidated by the Kabadiwalas who segregate the waste according to different types of polymers are further sold (25 INR/kg) to wholesalers or traders also called as aggregators. They are involved in compaction and bailing of the bottles that are then sold to the recyclers (30 INR/kg). The bottles are then subjected to manual sorting, labels, and caps removal. These are further washed and cut using specialized machinery. The end product is worth 55 INR/kg (http://www.petrecycling.in/). The recyclers use the same for either bottle production (closed loop) or fiber spinning (open loop) depending on the demand and process economics. The power, water, and cleaning chemicals consumption needed for the processing of the PET bottles at different stages in the downstream (gate to grave) life cycle is considered as the input.

8.2.5 Life Cycle Inventory Limitations and Uncertainties

Key assumptions made at the initial stages decide the extent and limitations of an LCA study. Data availability and data coherence pertaining to various stages of the life cycle amounts to an exhaustive assessment taking in consideration the complete picture related to the product or process. As assumptions of the LCA study are decided for the ease of practitioner many limitations are also translated due to the fixing these criteria. In the case of PET common limitations are regarding consolidated data collection throughout the life cycle from cradle to grave. Availability of region-specific datasets for the materials and energy flows involved at various stages of life cycle reduces the uncertainty involved in the impact results. Such parametric uncertainty is faced due to the input data obtained during life cycle inventory. Scenario uncertainties will include decisions such as inclusion and consideration of materials for labels and caps in the LCA study, filling of the product in the bottle and consumer usage stage of the life cycle or consideration of secondary packaging involved during the transportation and storage of the bottles. Model uncertainties are minimum as the production process for PET and the rest of the life cycle is simple and not involving a lot of complex mathematical modeling. The decision regarding the close and open loop recycle model establishes the value for the same.

8.3 Life Cycle Impact Assessment

8.3.1 Top Contributors

The impact categories considered during the selection of the impact assessment method to translate the inputs of the process in environmental burdens decide the orientation of the study. The categories related to land, water, air, resources, and human are given importance depending on the motive and context of the study. The top contributors are highlighted and give the crucial effect in the form of abstract concentration of chemicals that used to express the effect on the different sections of the ecology. Since the direct translation of materials in kg CO_2 emissions for the category of GWP (global warming potential) gives us an idea in terms of quantifiable impact. This impact along with other factors will express the complete extent of nuisance created in the environment. This nuisance value is either normalized to get an impact value in totality or at individual levels. The stage wise contribution of the impact can also be mapped using softwares or methods.

Mostly the production stage of PET resin pellet manufacture and PET bottle fabrication makes up the highest contribution for environmental burdens that is translated due to the usage of fossil fuels as utility and for the production of electricity. The emissions and the usage of fossil fuels are translated across various impact categories based on the assessment method chosen. Impact assessment methods express the results in terms of acidification, carcinogens, eutrophication, ecotoxicity, and smog, etc. being the top contributors for PET production [26].

Since the recycling of PET contributes and balances a part of emissions and burdens to the environment, higher percentage of recycle will improve the overall environmental performance of the product. Despite of high contribution due to the manufacturing stage a hefty part of the impact is offset by the recycling (closed or open) activity [35]. If the bottle is considered as complete waste and disposed off using conventional methods, landfill disposal will generate the maximum environmental impact [37].

8.3.2 Sensitivity Analysis

The later part of the LCA after generating the impact results is studying the dependability of the results related to specific parameters involved in the life cycle of the product and process. The extent of change relevant to a parameter can be quantified by subjecting the study to different values of the said parameter. Various scenarios based on real life situations or imaginary possibilities are implemented in order to find out the change in impact results. This exercise checks the robustness of an LCA study and also enables the practitioner to manipulate data to gain complete insight regarding each stage of the LCA. Since the production of the PET resin takes place at site A and the same is transported for injection and blow molding at sites B and C, transportation is a key parameter involved during the complete life cycle of the bottle. Various scenarios related to finding the optimum combination for transportation can be designed at the stage of sensitivity analysis. Energy sources can be switched from a fossil fuel based production to a renewable source such as wind or solar to observe the change in the overall impact of the process. Most efficient and optimum energy parameters related to machining of the polymer can be an assumed as a scenario to check savings that can be incurred with the change.

8.4 Qualitative Risk Screening of Selected Chemicals

Human toxicity and effect on the environment is the primary concern whenever chemicals are involved. The food and beverage packaging industry has already taken immense measures in order to assess the possibilities of leaching or migration of chemicals in the packed food products from the materials.

Antimony compounds are used as polycondensation catalysts and antimony trioxide used during the manufacture of PET covalently bonds with the PET matrix. Various studies pertaining to the leaching or migration of antimony trioxide to the bottled water and other beverages have been conducted [41,42]. Since it is used at a ppm level during production and bound by strong chemical bond after polymerization the risk of human toxicity reduces multifolds. Various other chemicals such as colorants, addititives, and stabilizers (phosphoric acid) involved in the PET production are thus subjected to such assessment frequently [2]. Ethylene glycol involved in the production do not pose a threat as it does not have the tendency to bio accumulate in humans and in the environment it is not persistent as it gets biodegraded or photo chemically oxidized readily [43]. Phthalate related compounds and thermal degradation compounds such as diethyl phthalate, formaldehyde, and acetaldehyde leaches at lower level and does not act as threat it being at a very low (ppb) concentration than the daily permissible limit of the same [44].

Since the product PET is inert in nature and is resistant to a list of chemicals the same has been translated in the tremendous usage of the same [45].

8.5 Conclusions and Suggestions

Such a study enables a holistic understanding of the complete life cycle of a product and in this case PET bottle, and insights related to each stage of the life cycle can be used to optimize the input–output material and energy flows. It also fosters policy development and critical and sustainable planning of material and energy flows taking environmental issues in consideration.

As seen the use of fossil fuels as utility at different stages in the life cycle of PET such as production and fabrication (gate to gate) and its use for the production of electricity that is consumed at different stages is a primary contributor to the environmental burdens. This has been translated across various impact categories. This source can be changed to a different renewable one, thus reducing the impact and making the product more sustainable and ecological.

Coherent and contextual data generation for specific process at different stages of the life cycle will decrease the uncertainty involved in the results. This assessment tool should be used generously to compare various system dynamics of the PET bottle at micro and macro levels at different stages, regions in the life cycle to gain better insight for sound decision-making. Since recycling of PET benefits tremendously depending on the quantum being collected, such processes should be fostered with better in place systems and incentivized policies.

References

[1] J. Fava, R. Dennison, B. Jones, M.A. Curran, B. Vigon, S. Selke, J. Barnum, in: J. Barnum (Ed.), A Technical Framework for Life-Cycle Assessment, Washington, DC, Report 1991.

[2] L.N. Ji, Study on preparation process and properties of polyethylene terephthalate (PET), Appl. Mech. Mater. 312 (2013) 406–410.

[3] I.J., Feldman Paula Global Packaging Landscape: Growth, Trends & Innovations, Virginia, Report 2015, pp. 1–20.

[4] L. Flanigan, F. Rolf, T. Montalbo, An Analysis of Life Cycle Assessment in Packaging for Food & Beverage Applications, Report 2013.

[5] Franklin Associates, Life Cycle Inventory of 100% Postconsumer HDPE and PET Recycled Resin From Postconsumer Containers and Packaging, Report 2011.

[6] S.K., Dogan Life cycle assessment of PET bottle (Thesis), Dokuz Eylül University, 2008.

[7] D.I., Nogueda Life cycle assessment of pet bottle recycling: a case study for Mexico (Thesis), Diplomatic Academy of Vienna, 2013.
[8] H. Thomas, K.B. Owen, Bevarage Container Review, Kamloops, British Columbia, Canada, Report 2013.
[9] PE International, Inc., Life Cycle Assessment of Polymers in an Automotive Assist Step, Boston, MA, Report 2012.
[10] Franklin Associates, Life Cycle Assessment of Drinking Water Systems: Bottle Water, Tap Water, and Home/Office Delivery Water, Oregon State, Report 2009.
[11] M. Bevilacqua, Integration of design for environmental concepts in product life cycle, in: M. Bevilacqua (Ed.), Design for Environment as a Tool for the Development of a Sustainable Supply Chain, Springer-Verlag, London, 2012, pp. 11–32.
[12] M.D. Tabone, J.J. Cregg, E.J. Beckman, A.E. Landls, Sustainability metrics: life cycle assessment and green design in polymers, Environ. Sci. Technol. 44 (21) (2010) 8264–8269.
[13] J.C.H. Hyun-Seob Song, A study on the comparison of the various waste management scenarios for PET bottles using the life-cycle assessment (LCA) methodology, Resour. Conserv. Recycl. 27 (1999) 267–284.
[14] J.M.L. Reis, R. Chianelli-Junior, J.L. Cardoso, F.J.V. Marinho, Effect of recycled PET in the fracture mechanics of polymer mortar, Constr. Build. Mater. 25 (6) (2011) 2799–2804.
[15] E. Hochschorner, G. Finnveden, Evaluation of two simplified life cycle assessment methods, Int. J. Life Cycle Assess. 8 (Graedel 1998) (2003) 119–128.
[16] M.C. McManus, C.M. Taylor, A. Mohr, C. Whittaker, C.D. Scown, A.L. Borrion, et al., Challenge clusters facing LCA in environmental decision-making—what we can learn from biofuels, Int. J. Life Cycle Assess. 20 (2015) 1399–1414.
[17] M., Schmidt, A. Ostermayer , D. Bevers, Life cycle assessment of PET (polyethylene terephthalate) bottles and other packaging alternatives, 2000.
[18] Franklin Associates, LCI Summary for PLA and PET 12-Ounce Water Bottles, Kansas, Report 2007.
[19] Rajendra Kumar Foolmaun, Ramjeawon Toolseeram, Life cycle assessment (LCA) of PET bottles and comparative LCA of three disposal options in Mauritius, Int. J. Environ. Waste Manag. 2 (1/2) (2008) 125–138.
[20] Franklin Associates, Life Cycle Inventory of Three Single-Serving Soft Drink Containers, Kansas, Report 2009.
[21] J.M. Fry, B. Hartlin, E. Wallén, S. Aumônier, Life Cycle Assessment of Example Packaging Systems for Milk, Report 2010.
[22] PE-Americas and Five International, Life Cycle Assessment Summary, Report 2010.

[23] B. Kuczenski, R. Geyer, Life Cycle Assessment of Polyethylene Terephthalate (PET) Beverage Bottles Consumed in the State of California, California, Report 2011.
[24] C. Belley, Final Assessment Report Comparative Life Cycle Assessment Report of Food Packaging Products, Montreal, Quebec, Report 2011.
[25] W.P. Flanagan, H. Dhaliwal, M. Browne, Glass or plastic: an environmental life cycle assessment (LCA) and related economic impact of contrast media packaging, ECR 2015, European Society of Radiology, New York, 2015, pp. 1–16.
[26] Kang, D. Environmental Evaluation of Non-Alcoholic Single-Serve PET Beverage Bottles in the State of California Using Life Cycle Assessment and System Dynamics, Michigan State University, Report 2015.
[27] ECO Paper Loop, LCA of Selected Packaging Products, Report, 2015.
[28] A.K. Singh, H. Agarwal, R. Sinha, S.K. Jha, A. Prakash, Life cycle assessment of polyethylene terepthalate (PET) bottles, ELK Asia Pac. J. (special issue), Proc. Int. Conf. Adv. Recent Innov. Mech. Prod. Ind. Eng., 2006.
[29] B. Kuczenski, R. Geyer, PET bottle reverse logistics—environmental performance of California's CRV program, Int. J. Life Cycle Assess. 18 (2) (2013) 456–471.
[30] P. Europe, Polyethylene Terephthalate (PET) Bottle Grade, Belgium, Report 2008.
[31] L. Shen, E. Nieuwlaar, E. Worrell, Life cycle energy and GHG emissions of PET recycling: change-oriented effects, Int. J. Life Cycle Assess. 16 (2011) 522–536.
[32] L. Chen, R.E.O. Pelton, T.M. Smith, Comparative life cycle assessment of fossil and bio-based polyethylene terephthalate (PET) bottles, J. Clean. Prod. 137 (2016) 667–676.
[33] M. Garfí, E. Cadena, D. Sanchez-ramos, I. Ferrer, Life cycle assessment of drinking water: comparing conventional water treatment, reverse osmosis and mineral water in glass and plastic bottles, J. Clean. Prod. 137 (2016) 997–1003.
[34] F. Gironi, V. Piemonte, Life cycle assessment of polylactic acid and polyethylene terephthalate bottles for drinking water, Am. Inst. Chem. Eng. 30 (3) (2011) 459–468.
[35] L. Shen, E. Worrell, M.K. Patel, Open-loop recycling: a LCA case study of PET bottle-to-fibre recycling, Resour. Conserv. Recycl. 55 (1) (2010) 34–52.
[36] H.L. Raadal, O.M.K. Iversen, I.S. Modahl, LCA of beverage container production, collection and treatment systems, Krakeroy, 2016.
[37] J. Hopewell, R. Dvorak, E. Kosior, Plastics recycling: challenges and opportunities, Philos. Trans. R. Soc. B 364 (2009) 2115–2126.

[38] H. Song, K. Moon, J.C. Hyun, A life-cycle assessment (LCA) study on the various recycle routes of pet bottles, Korean J. Chem. Eng. 16 (2) (1999) 202–207.

[39] ISO, ISO 14040:2006 Life Cycle Assessment—Principles and Framework, International Organization for Standardization, Report 2006b.

[40] ISO, ISO 14044:2006 Life Cycle Assessment—Requirements and Guidelines, International Organization for Standardization, Report 2006c.

[41] S. Rungchang, P. Suthiluk, S. Numthuam, T. Satake, Antimony Leaching From Polyethylene Terephthalate (PET) Plastics Used for Beverage in Japan, Ibaraki, Japan, Report 2011.

[42] International Antimony Association, Studies Confirm the Safety of the Use of Antimony Compounds for PET Bottles, Brussels, Belgium, Report 2014.

[43] R. Gomes, R. Liteplo, M.E. Meek, Ethylene Glycol: Human Health Aspects, Geneva, Report 2002.

[44] C. Bach, X. Dauchy, M. Chagnon, S. Etienne, C. Bach, X. Dauchy, et al., Chemical migration in drinking water stored in polyethylene terephthalate (PET) bottles: a source of controversy, Water Res. Elsevier 46 (3) (2012) 571–583.

[45] Plastics Europe, General Chemical Resistance of PET—Products, Brussels, Belgium, Report 2015.

9 Applications of Waste Poly (Ethylene Terephthalate) Bottles

Ajay Rane[1], A.R. Ajitha[3], M.K. Aswathi[3], P. Manju[4], Krishnan Kanny[1] and Sabu Thomas[2,3]

[1]Composite Research Group, Department of Mechanical Engineering, Durban University of Technology, Durban, South Africa, [2]School of Chemical Sciences, Mahatma Gandhi University, Kottayam, Kerala, India, [3]International and Inter University Centre for Nanoscience and Nanotechnology, Mahatma Gandhi University, Kottayam, Kerala, India, [4]Department of Plastics Technology, Central Institute of Plastics Engineering and Technology, Chennai, Tamil Nadu, India

9.1 Introduction

Chapter 1, PET Chemistry, in the book provides details on poly(ethylene terephthalate) (PET) chemistry, but to be in short—esterification between terephthalic acid and ethylene glycol (EG), with water as a by-product also transesterification reaction between terephthalic acid and dimethyl terephthalate in the presence of methanol forms PET monomer. PET is a thermoplastic used in manufacturing of bottles, films, fibers and fabrics, high-quality carpets, packaging for detergents, cosmetic, foils, car spare parts, pillow fillings for allergic persons, and other molded items for household and industrial applications. Existence of PET in amorphous and crystalline forms depends on its thermal history and their processing conditions. Amorphous PET is transparent, and as white and opaque for crystalline form. It is rightly mentioned that with increasing population, it is difficult to regulate huge mass of residual materials generated from enormous industrial activities. Wastes are those residues which are not recycled, reused, reclaimed, and these wastes which do not fit in environment become hazardous and detoriate the quality of environment and add to pollution. Recycling through available options (via mechanical, chemical, and biological routes) has become a topic of research in industries and academics. Mechanical recycling suits best to thermoplastic materials, i.e., for a polymer matrix which can be reprocessed and remelted into different forms of products via any of the melt mixing techniques (e.g., injection and extrusion molding). However, thermosets cannot be reprocessed like

Recycling of Polyethylene Terephthalate Bottles.
DOI: https://doi.org/10.1016/B978-0-12-811361-5.00009-2

thermoplastics, but chemical recycling is used to recycle them to feedstock or can be used as a carrier (cement kilns). Plastics have contributed to development and progress of society through spreading its roots deep within applications covering depth of sea to land to space including aerospace, medical, household, etc. We can find plastics everywhere from clothing to shelter, transportation to communication, entertainment to health care, hence they play a key role in human behavior and are a part of human beings, due to light weight, high strength, and easy processing methods. Most of the bottles used for single serves (water, soft drinks, and juices) are made of PET as pointed out in manuscript by Dutta et al., according to NIIR Project Consultancy Services. Consumption of these PET single serve bottles is increasing worldwide, hence disposing these bottles has become a major concern for environmentalist and organizations working in areas related to waste reduction and minimization. Major part of PET bottles ends up in landfills and incinerators; remaining portion is mechanically grinded into powders. This chapter discusses in detail about applications of mechanically and chemically recycled PET (r-PET) into powders to blend with other polymers and starting ingredients for synthesis of many other polymers, respectively. Virginija through her editorial write up in Environmental Research, Engineering and Management mentions that plastic recycling is very important for at least two main reasons—first, to reduce the increasing volumes of plastic wastes and second, to generate value-added materials [1,2]. This chapter concerns major applications of waste plastic bottles, as depicted in Fig. 9.1.

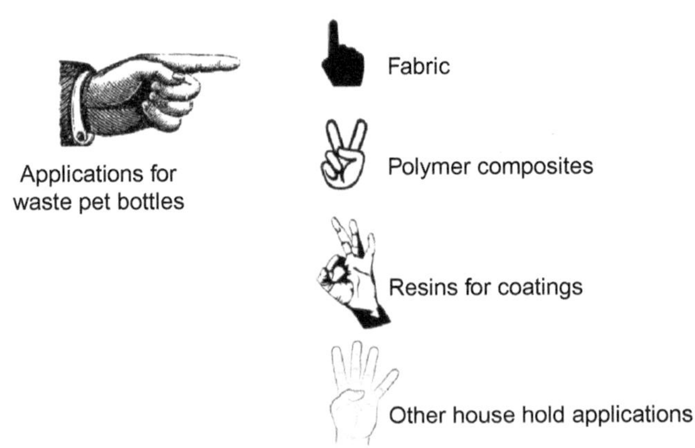

Figure 9.1 Major applications of waste PET bottles.

9.2 PET Bottles–Fiber–Fabric

The London Olympics "2012" has a major role in commercializing the recycling possibilities of different polymers and the efficient application of those in different fields. Out of those, the athletic wear would have attracted the most, because the clothings and accessories worn by the athletes symbolized sustainability as they were manufactured from recycled materials. The uniforms that US men's basketball team wore were made from r-PET bottles. From 2015, Indian cricket team also started using jerseys made by Nike from r-PET bottles. These milestones point toward the recycling possibilities of PET bottles into fiber and further into a fabric. As the major application of polyester goes for bottles and textiles, bottles can be effectively made into textiles.

Before going further into the topic, some terminologies related to textile technology are introduced. *Spinning* is the process for producing filaments/fibers; *melt spinning* is the process for producing filaments/fibers for polymers that can be melted without undergoing thermal degradation; *fiber* is a raw material for making textile products, with its length more than that of its diameter; *staple fiber* is a fiber with fixed length and diameter (Fig. 9.2); *filament* is a continuous strand of fiber with longer length than that of staple fiber (Fig. 9.2); *yarn* is a long-twisted strand of bundles of fibers, which is the end product of melt spinning process; *thread* is two or more yarns twisted together, forms thread; *fabric* is a cloth produced by weaving or knitting process using textile threads or fabric.

Polyester fiber is the most commonly used material for textile application. It is most widely chosen because of its advantages such as flexibility, light weight, reduced wind, and drag resisting ability. The other characteristics of PET fibers are resistance to stretching and shrinking, wrinkle free, quick drying, resistance to abrasion, etc. [3]. The polyester

Figure 9.2 Pictorial representations of filament and staple fiber.

fibers are made by melt spinning process in which the PET granules are melted and extruded through the spinneret, wherein the polymer melt passes through a metering pump to control and filter the flow. The quench air cools the extrudate filaments and the obtained bundles of continuous filaments are called as tow. On further processing of tow, such as drawing, crimping, spin finish applications, these tows are cut into fixed lengths. These cut fibers are called polyester staple fibers (PSF) (Fig. 9.2). Drawing is the forced pulling out of tows for further processing. Crimping helps in increasing the inter-fiber adhesion resulting in increased cohesive forces between them. Spin finishing technology ensures the smoother functioning of fibers during the spinning process [4–8].

The color of PET fiber is slightly off-white or translucent (depending on whether it is crystalline or amorphous in nature) as mentioned earlier in Section 9.1. However, the polymer system is extremely crystalline in nature with 65%–85% crystalline regions and 35%–15% amorphous regions. Most of its physical and chemical properties arise due to this microstructure. The important physical and chemical properties of PET fibers are given in Fig. 9.3. The diameter of PSF is usually varied according to the end-use application, even though the general diameter ranges between 12 and 25 μm and the aspect ratio (fiber length to breadth ratio) exceeds 2000:1 [5]. These PSF are spun to form yarn. This spun yarn is used in textiles, bedspreads, pillow covers, sportswear, athletic shoes, etc.

The waste PET bottles are collected and sorted manually or automatically at the recycling centers and bundled to large bales. These bales are taken into a reclaiming center where the bottles are thoroughly

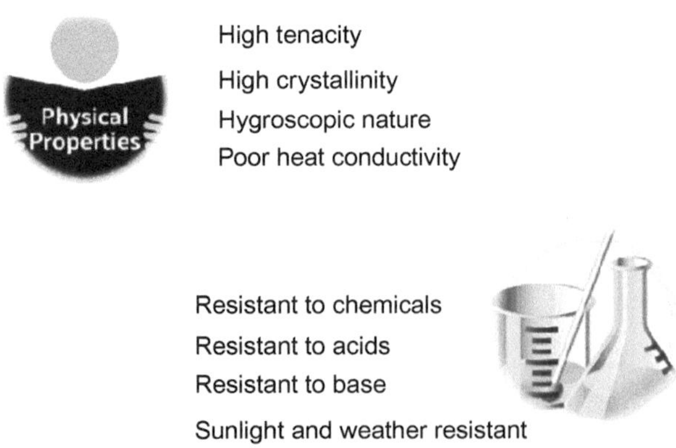

Figure 9.3 Physical and chemical properties of PET fiber.

cleaned with removal of labels, caps, and rings and separated according to their color. This is because the colorless bottles yield whitish yarn, whereas colored bottles give the colored yarns. After cleaning, bottles are shredded into PET flakes. The flakes are then washed again in a sterilizing bath. The light colored flakes are bleached to obtain the colorless yarn, while the flakes from the dark colored bottles are dyed with a dark color. Then as similar to that of virgin PET granules, the PET flakes are melt-extruded and form filaments that are five times finer than human hair. Then these filaments are sent over hot metal rollers to stretch and realign the PET molecules. This brings crystallinity to the PET fibers and is cut into r-PSF. Then it is sent to spooling machine where the yarns are spooled. Further this yarn is ready to be made into PET fabrics [4–12]. All the processes are similar to that of virgin PET granules except the extra cleaning procedures adopted for the PET flakes. Fig. 9.4 shows the procedure for obtaining the fibers from PET bottles. Rather than PSF, there are many other categories where the recycled fibers get in through. Fiber fill is an application where the short fibers are used for the filling of pillows, cushions, etc. Partially oriented yarn is also a PET yarn used for making fabrics [13]. PET fibers are used in geotextiles, which are permeable fabrics which when

Figure 9.4 Converting waste PET bottles to fabrics.

used with soil have the capability to filter, reinforce, drain, and protect the soil used during road construction [14]. For any of the manufacturing of fabrics from r-PET fibers, the main criterion is that the properties should meet with the virgin PET fibers. Thus, the garments once produced from r-PET can be again recycled and reused and thus it enters a closed-loop system.

Most of the leading textile companies have now extended their technologies in developing fibers from recycled materials with a perspective of making green and sustainable fabrics. Also the effective utilization of waste materials, which usually goes for landfilling or incineration process, can reduce the negative impact of them in the environment. And some of the manufacturers incorporate the recycled content with the virgin PET granules to make their fabrics.

Repreve is a brand of recycled fiber produced by Unifi that is made from r-PET including postconsumer bottles and postindustrial waste from manufacturing wastes. It is used in numerous brands including Quicksilver, Haggar clothing, Adidas, Russell, Patagonia, etc. Through this, they could convert 630 million PET bottles from the landfilling to useful fibers. For making Denali Jackets alone, yearly they were using over 30 million waste PET bottles. Ecocircle, by Tenjin, is also a brand of fiber made from r-PET including PET bottles. These are used in making men's and women's wear. As of now around 150 companies are involved in this group. Camtex fabrics has introduced fabrics with recycled content mainly derived from recycled bottles and were used in the manufacture of Earth keeper shoes, boots, and clothings. In 2005, Chung Shing Textile Company started manufacturing high-quality PET fiber from PET bottles and branded them as GreenPlus and is being used by many manufacturers for making different textile products. From 2010, Nike was using discarded PET bottles from the landfill areas of Japan and Taiwan to produce r-PET fiber to make jerseys for national soccer team for Brazil, the Netherlands, Portugal, Serbia, etc. Hanes Ecosmart was using r-PET from bottles to make their fabrics. Patagonia, the first manufacturer to use waste materials to transform to clothes, started using waste scrap from PET soda bottles to make clothings. In 2012, Levis introduced a new brand of jeans Waste Less in which each pair of jean has r-PET fiber from an average of from 8 to 12 PET bottles. From 2014 onwards, Cone Denim together with Unifi started using r-PET fibers in their jeans Cone Touch and a pair of jeans uses an average of eight r-PET bottles [15,16].

This commercialization of r-PET fibers helps to spread the knowledge easily and helps in building a sustainable environment. This also encourages people to probe into recycling possibilities of different polymers into different fields with overcoming all the present limitations of recycling technologies. Also, the technologies need to be addressed to everyone so that many entrepreneurs can be attracted toward recycling and help toward solving the problem related to waste PET bottles.

9.3 PET Bottles—Resins for Coatings and Recycled Polymer Composites

Chemical recycling processes—glycolysis, aminolysis, ammonolysis, hydrolysis, and methanolysis—and combination of the above-mentioned chemical process are discussed in detail in initial chapters of this book. The oligomeric products—hydroxyl and amine terminated oligomers, polyols—are products of these chemical process. These recycled oligomeric products are further reacted with suitable cross-linkers and hardeners to synthesize resins for coatings and composites. Figs. 9.5 and 9.6 show the process involved in converting PET bottles into resins for coatings and recycled polymer composites, further pigments, and other essential additives are added to have a desirable property for coating and composite material.

Figure 9.5 Converting waste PET bottles to resins for coatings.

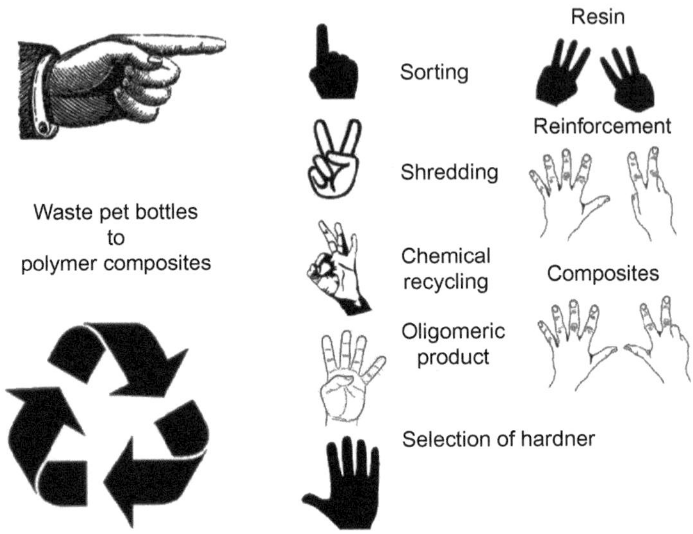

Figure 9.6 Converting waste PET bottles to resins for polymer composites.

Recycling of PET can be done through either physical methods (mechanical methods) or chemical methods. Physically r-PET is generally used to make blends with other polymers, while the chemically r-PET is introduced as starting ingredients for synthesis of many other polymers [17]. One of the major advantages of physical recycling is that the basic polymer is not altered during the process. Compared to chemical recycling methods, physical recycling of PET shows the following advantages such as simplicity, flexibility of feedstock supply, low coast, low negative environment effect, and no need of sophisticated equipment. The major disadvantages of physical recycling are decrease of product properties with every cycle and reduction in melt viscosity or the reduction of the molecular weight caused by thermal and hydrolytic degradation. Also during melt processing of PET, cyclic and linear oligomers are produced which negatively affect the properties such as printability and dyeability of final product [18]. The heterogeneity of the solid waste is another main issue faced in the mechanical recycling, because mechanical recycling becomes difficult as the complexity and contamination of the solid waste increases. During thermal processing, the acid components produced from the contaminants catalyze the hydrolysis of the PET's ester linkages. Akçaözoğlu et al. [19] used granulated waste PET bottles as an aggregate in the production of structural light weight concrete. They prepared mortar samples using PET aggregates and sand aggregates together. The mortars with PET and sand aggregates show better properties than mortars with PET aggregates alone. Akçaözoğlu and Atis [20] prepared light weight mortar using waste PET aggregates, granulated blast furnace slag, and fly ash. Frigione prepared concrete with fine aggregates of waste unwashed PET bottles [21]. Hassani et al. [22] used PET aggregates in asphalt concrete. They used 3 mm diameter granules of PET in order to reduce the environmental problem of PET disposal.

In chemical recycling, PET undergoes either total depolymerization or partial depolymerization and gives monomers and oligomers, respectively. Chemical recycling of PET can be done through hydrolysis, methanolysis, glycolysis, aminolysis, ammonolysis, etc. In chemical recycling, the polymer chains undergo transformation and degrade into monomer units. Products from chemically r-PET can be used to produce polyurethane and unsaturated polyester (UP) resin. It was found that polyurethane has many applications in insulation, seating material, and artificial leather. The polyester polyols obtained from the PET glycolysis and aliphatic diacids can be used as starting material for polyurethane synthesis. Rusmirovic et al. synthesized the alkyd resins from waste PET bottles through glycolysis process using trimethylolpropane and trimethylol ethane in the presence of tetrabutyl titanate catalyst. The glycolizate product obtained from the PET waste bottles are applied in the alkyd resin production. Thereafter they prepared nanocomposites using alkyd resins and methacrylol modified silica particles. The obtained nanocomposite coatings exhibited good mechanical and anticorrosive properties [23]. Pogrybnyak et al. developed a process for producing polymer composites based on PET and wood flour. The resin obtained after the partial glycolysis of recycle PET is used to prepare polymer composites with wood flour. Then, the resulted wood filled composites were pressed to produce boards, panels, trays, and sheets [24]. Glycolysis of PET flakes can be done using EG, propylene glycol (PG), diethylene glycol (DEG), and triethylene glycol in the presence of zinc acetate catalyst. The glycolysis product of PET, maleic anhydride, and PG are reacted together to form UP resin which is further reinforced with particulate and fiber like fillers to fabricate composite materials. Ahmed et al. have prepared composites using rice husks and UP resin obtained from the glycolysis of waste PET [25]. Glycolysis of PET shows the advantages like simplicity, flexibility, low capital costs, and usability of conventional PET production plants. But it also shows some limitations that it cannot produce clear virgin PET [26]. Tawfik and Eskande prepared polymer concrete (PC) using marble waste and styrenated polyester where the polyester is the glycolysis product of

PET bottles [27]. Methanolysis of PET bottles produce Dimethyl terephthalate (DMT) and EG. Compared to glycolysis process, in methanolysis the PET feedstocks with lower quality are also acceptable. The products of methanolysis, EG, and methanol are easily recycled and recovered. Even though methanolysis has high processing costs, the relatively low feedstock costs of this compensate for its high processing costs [26]. The main drawback of hydrolysis is the need of high temperatures, high pressure, and long-time needed for complete depolymerization [28]. Pusztaszeri et al. proposed acid hydrolysis to recycle PET using concentrated sulfuric acid to avoid high temperature and pressure conditions used in hydrolysis [29]. But it also has some drawbacks that the cost of this process is high because it needs to recycle the large amounts of concentrated H_2SO_4 and it also needs to purify EG from acid [30]. Tawfik et al. used the aminolysis product obtained from the scrap PET bottles to form anticorrosive paints for the protection of steel structures [31]. Spychaj et al. reported the use of aminolysis and aminoglycolysis product of waste PET as hardener in liquid epoxy resin and epoxy resin which can have applications in adhesives, coatings, and insulation field [32]. Zang et al. reported the aminolysis of waste PET using ethanolamine. The product obtained from the aminolysis, bis (2-hydroxyethyl)terephthalamide has the potential to undergo further reactions to yield secondary value-added product, such as UPs, polyurethanes, epoxy resin hardeners, and non-ionic polymer surfactants [33]. r-PET bottles can be used to fabricate composite panels which can have applications in construction field. Also recycled waste PET bottles are used as voluminous filler material in composite panel. In these cases, the r-PET material is added in a fragmented state. Recycled waste PET bottles also found applications in asphalt, PET char as additive in composites, PET particles in polyurethane foams and as a PC [34]. Kamall and Rizvi made PET composite panels using the whole PET bottle (including the caps) without affecting its physical properties. The prepared composite panels are capable to reduce the PET bottle waste and the air pollution. It also helps to reduce the cost of energy used for heating and cooling purposes of the buildings. By using these composite panels, thermal insulation of buildings is possible, and importation of expensive insulation materials and its transportation are also reduced [35]. PET bottles are also used as voluminous filler materials in composites. Among other used filler insulation materials such as phenolic foam, rigid polyurethane, mineral fiber, rigid polyisocyanurate, extruded polystyrene, and expanded polystyrene, the PET bottles provide excellent thermal insulation because of entrapped air in the empty PET bottles [35]. Nováková et al. produced a special kind of brick from r-PET material which has been found application as seating objects [36]. The recycling of PET bottles was done through different steps (Fig. 9.7A). From the recycled material, the PET brick was produced using blow molding technique. The prepared PET bricks show an interlocking capability which reduces or eliminates the use of binding material of bricks. The structure of PET brick from different angle is shown in Fig. 9.7C. These bricks are used to make seats as shown in Fig. 9.7B.

Mahdi et al. prepared PC from r-PET bottles. This PC can be used instead of ordinary Portland cement. The UP resin produced from the glycolysis of r-PET waste was used as a binder to generate PC [37]. Another group used PET char obtained from the pyrolysis of PET waste to prepare epoxy composite. Up to 50 wt.% of char addition is possible to this system. The mechanical, electrical, and hardness test of these composites with different wt.% of char at different temperatures shows good values [38]. The gas and liquids produced during the pyrolysis are used as fuel and chemical feedstocks and the remaining char is used to produce composite material so thereby reduces the soil pollution by not depositing the char in the land. For the organic oxidation of dyes Bento et al. fabricated red mud-PET composites. The prepared composite is applicable for the oxidation of methylene blue dye. They prepared the composite with different wt.% of PET powder such as 10, 15, and 20 wt.%. The composite with 15 wt.% shows highest methylene blue removal because it contains highest amount of reducible Fe^{3+} ions. The carbon from the PET part reduces Fe^{3+} to Fe^{2+} ions. This composite

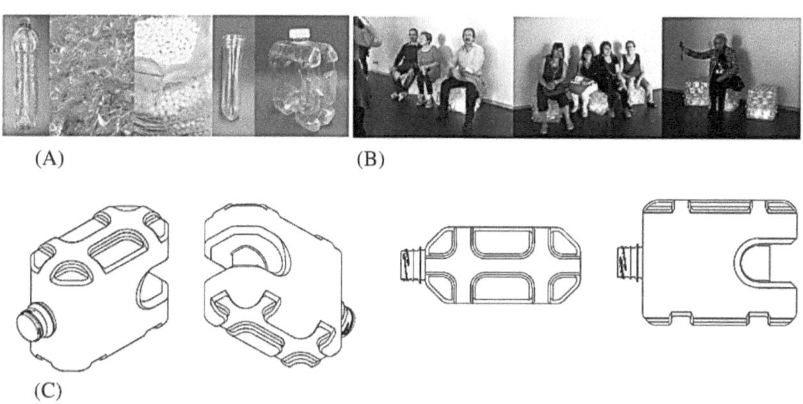

Figure 9.7 Converting waste PET bottles to PET bricks.

fabrication of red mud and PET powder reduces the solid waste disposal and liquid waste problems [39].

Çınar and Kar fabricated composite materials using particles obtained from PET waste bottles and marble dust [40]. The prepared composite shows a synergic effect of both marble dust and PET. The properties of composites are higher than that of pure PET waste. The marble piece shows good interaction with PET particles. This PET marble dust interaction and non-flammability properties of marble pieces lead to a construction material with good quality in low cost.

9.4 Coating Applications

The chemically recycled products of PET bottles have a wide application in coating industry, since it is a source for the raw materials for the preparation of coating materials in less expensive manner and in an environmental point of view. Generally, the recycled products of PET bottles were used for the preparation of saturated and UP resins, alkyd resins, polyurethane dispersions (PUD), melamine formaldehyde resins, textile dyes, softeners in textile finishing, and binders for paints, etc., with similar properties of ordinary ones prepared from its corresponding monomers. Thus, it can be said that the r-PET products have significance in various fields such as in ships, containers, automotives, paints, UV curable coatings, powder coatings, refineries, marine equipment, chemical plants, and in textile industries [41].

Chemical recycling methods such as aminolysis, hydrolysis, and glycolysis are the most commonly used methods for the PET bottle recycling; furthermore, the recycled product is characterized and functional groups are determined by structural analysis, later depending on properties of coating required (external coatings, interior coatings, and special purpose coatings), different additives are added into recycled product along with the suitable cross-linker to synthesize saturated and UP resins, alkyd resins, alkyd-amino resins, PUD, melamine formaldehyde resins, etc., which have various applications in coating industry. Later its performance properties are evaluated and compared with conventional ones.

In the coating industry, the additives have very much importance, for example, paints contain different additives and pigments. Generally, the dispersion of these pigments and additives in a continuous phase gives the paint in its applicable form. This continuous phase has a significant importance in the preparation of good coatings and is known as binders.

The recycled products of PET bottles can act as a good binder for paints with required properties. Nowadays, coating industry looks into the development of water-based resins to reduce the usage of volatile compounds due to the increased attention to the healthy environment [42]. Waste PET bottles were used for the preparation of various polymeric coating binders in industry level. Such various binders also include the alkyd-amino resins, alkyd resins, polyurethane resins, epoxy resins, and UP resins.

9.4.1 Epoxy Resins

Epoxy resins are the important thermoset polymers and have wide applications as adhesives and coatings. Increased functionality will improve the adhesion of coatings and have wide applications in steel coatings in various fields of refineries, chemical plants, marine equipment, etc. These epoxy resins can also be applicable in paints as binders. Bal et al. synthesized epoxy-based paints from recycled products of PET bottles and they observed that it can be used as a binder in paint [43].

Literature confirms polyester resins of great importance in powder coating applications. The glass transition temperature of a polyester resin used in powder coating must be higher than the room temperature to prevent blocking phenomena, so a polyester resin must contain large amounts of terephthalic acid. r-PET contains large amounts of terephthalic acid, thus it can suggest that r-PET is a promising candidate as a raw material for the polyester resins and thereby for powder coatings [44].

9.4.2 Polyurethane Dispersions

PUD are binary colloidal systems formed by the particle dispersions in an aqueous phase. They are good candidates for coatings due to its very low organic content, easily handling property, and wide range of film hardness. PUD have a wide application in polyurethane coating industry on considering its versatility and superior properties such as abrasion resistance, better adhesion to many polymers, and low temperature flexibility [45].

Polyurethanes are fast developing products of the coating industry with outstanding performance. But they are in poor availability due to the high cost of raw materials. Thus the usage of r-PET in the synthesis of polyurethane will bring down their cost without compromising performance. Patel et al. used the oligoesters obtained on the depolymerization of PET for the synthesis of polyester polyols. These polyols were used to prepare two-pack polyurethane coating systems [46].

9.4.3 Alkyd Resins

Alkyd resins are the most widely used polymers for paints and other coating applications with low volatile organic compounds. These are highly branched polyester composed of polyhydric alcohols, dibasic acids, and higher fatty acids. Alkyd resins with improved bending impact, gloss values, fast drying hardness, and good adhesion can be prepared from the recycled products of PET bottles in an economic way with eco-friendly manner. And they are widely used in ambient temperature curing coatings. Recycled products of PET are highly efficient for the production of alkyd resins by the controlled usage of r-PET; one can overcome the disadvantage (turbidity nature) of the ordinary alkyd resins prepared from ordinary terephthalic acid (t-PA). The turbidity of the alkyd reins prepared from t-PA is due to its high crystallinity and are not compatible with the other components [44]. The alkyd resins have more applications than the other recycled products of PET bottles due to its low viscosity and fast drying property. It can be used for the preparation of high-gloss decorative paints and also for carbon steel as corrosion protective coating [44,47]. Kawamura et al. observed that the polyester resin synthesized from r-PET have structure same as that of conventional polyester synthesized by the ordinary method. And the film properties of powder coatings formulated with polyester synthesized from r-PET instead of EG and t-PA were also comparable to those of a conventional coating. They also successfully synthesized alkyd resin having the same characteristics as a conventional resin from r-PET instead of EG and PA by modifying its monomer composition and the reaction. Later the ambient temperature curing coatings formulated with this alkyd resin showed comparable properties to those of conventional resin [44]. Kathalewar et al. synthesized polyester polyol by polyesterification of glycolyzed monomeric product and then PU coatings were prepared using this polyester polyol and various kinds of polyisocyanate curing agents. And optical, mechanical, chemical, and thermal properties of PU coated films were examined [48]. Kawamura et al. synthesized polyesters and alkyd resins for powder coating and for ambient temperature coating, respectively, from PET recycled products. They attained same structure and characteristics as the conventional ones prepared from the ordinary monomers such as EG and terephthalic acid [44]. Jamdar et al. were successfully prepared polyurethane coatings from the polyester polyol (obtained from the recycled material of PET—bis(2-hydroxyethylterephthalate) (BHET)—in combination with bio-based monomers) and various commercial polyisocyanate curing agents. Performance properties and thermal properties were evaluated [49]. Torlakoğlu and Güçlü [50] prepared alkyd-amino resins from the PET glycolyzed products for coating applications. The alkyd amines obtained from PET recycling were blended with UF, MF, and UF/MF (1:1).

Preprocess involved in collection of waste PET bottles is one of the challenges. Public awareness regarding disposal methods is necessary. Finally, to conclude trails in purification of recycled product obtained from chemical recycling is still going so that pure form of the resins could be obtained and made use of speciality applications. Challenges in blending shredded waste PET flakes need to be overcome by looking for available surface treatment process to increase the compatibility between polymer matrix and waste PET bottle flakes, to fabricate high performance composites for structural and functional purposes. Fabrics obtained from waste PET bottles are successfully used in manufacturing of clothing as discussed earlier.

9.5 Microfibrillar Polymer Composites

Development of high performance micro- and nanofibrillar composites from waste PET bottles and polyolefin bags is one of the practices in our group. Microfibrillar composites (MFCs) are a novel group of fiber-reinforced composites. Manufacturing of in-situ MFCs provides a potential route to enhance properties of all-purpose thermoplastics (such as LDPE, HDPE, and PP), using engineering thermoplastics (such as PET, polyamide, polycarbonates, and polybutylene terephthalate) [51–59]. MFCs are dissimilar from the classical composites and conventional blends as microfibrillar structure of the reinforcing polymer is incorporated within the polymer matrix during processing. The general scheme of the experimental setup used for the MFC preparation is given in Fig. 9.8.

Four elemental steps are envisaged: (1) Melt blending of immiscible polymeric components whose apparent melting temperatures (T_m) differ by at least 30°C (the T_m of the reinforced minor phase is greater); (2) extrusion of the as-prepared blends; (3) drawing the obtained extrudates to fibrillate the minor, dispersed phase; and (4) annealing or isotropization of the as-drawn extrudates above the T_m of the major phase. In count to the

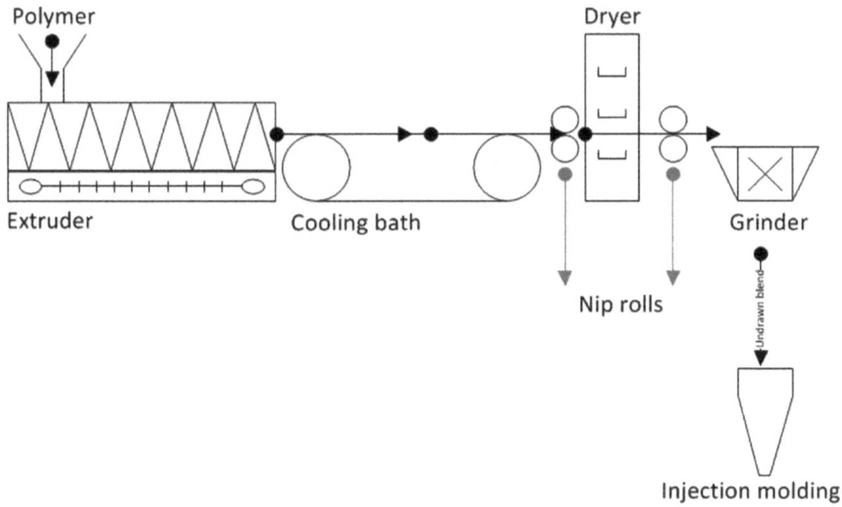

Figure 9.8 Scheme of the experimental setup for microfibrillar in situ composites preparation [55].

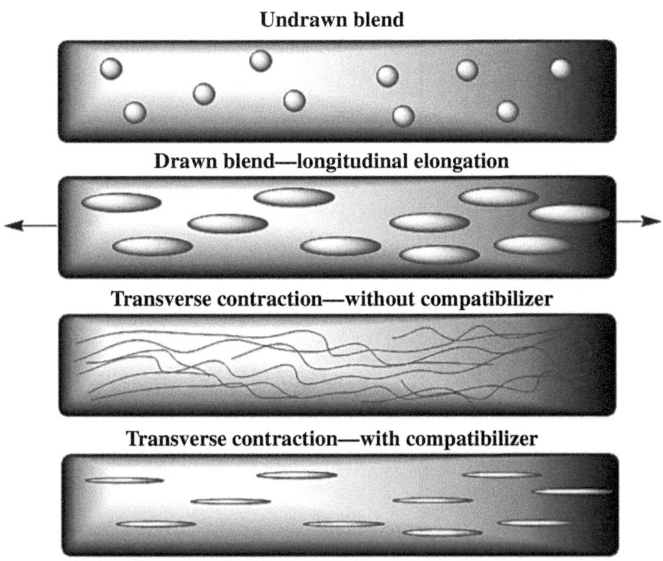

Figure 9.9 Mechanism for formation of microfibrils in polymer blends during cold drawing [60].

synergistic property of the resulting mechanical properties, MFCs offer another important advantage wherein the reinforcing element is also a conventional thermoplastic. Thus no mineral additives are involved [54].

Fig. 9.9 illustrates the conversion of the dispersed spheres into microfibrils due to coalescence during the cold drawing [60]. Elongation of the

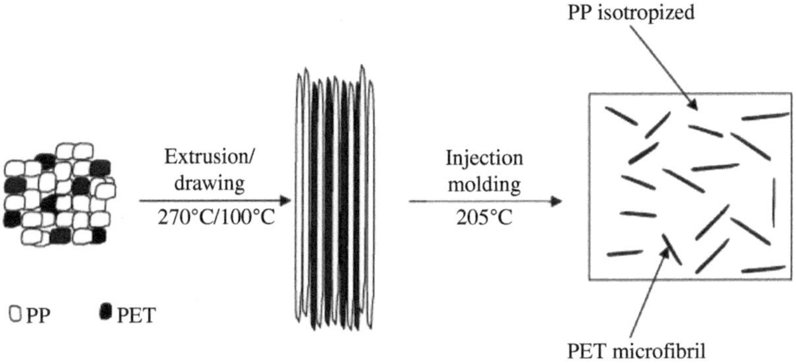

Figure 9.10 Morphology development scheme of the blend during extrusion and injection molding [55].

dispersed spheres into rotational ellipsoids or fibrillar particles with lengths corresponding to the draw ratio applied. With the progress of the stretching process, these elongating particles can come into contact with each other due to their irregular movements within the matrix flow, and can eventually merge to form long "endless" microfibrils. The contribution of coalescence in the formation of long microfibrils can explain the existence of short fibrils. The morphology develop is illustrated in Fig. 9.10. After extrusion and drawing, both the phases involved attain fibrillar morphology.

To conclude, waste PET in the form of packaging and structural products should not be considered as harmful to environment, but should be treated as a resource with major advantages. Voluminous applications and no proper method for recycling have created social concern regarding PET waste management. There is no scientific standing by simply banning one of the most important engineering thermoplastic. By using effective recycling techniques, waste PET could be transformed into a value-added material. Simply banning plastics shall affect the entire human habits, as human beings are becoming more dependent on plastics for all needs. Additionally, plastic industries provide employment to major part of population.

References

[1] V. Jankauskaite, Recycled polyethylene terephthalate waste for different application solutions, Environ. Res. Eng. Manag. (2016) 5–7. Available from: https://doi.org/10.5755/j01.erem.72.1.15260.

[2] M.B. Sushovan Dutta, J.N. Nadaf, Mandal: an overview on the use of waste plastic bottles and flyash in civil engineering applications, Proc. Environ. Sci. 35 (2016) 681–691.

[3] https://www.ivc-ev.de/live/index.php?page_id = 72.
[4] A. Telli, N. Özdil, Effect of recycled PET fibers on the performance properties of knitted fabrics, J. Eng. Fabrics Fibers (JEFF) 10 (2) (2015).
[5] http://nptel.ac.in/courses/116102026/38.
[6] http://www.fibre2fashion.com/industry-article/4393/problems-and-their-remedies-in-polyester-processing-part-iii.
[7] K.K. Leonas, The use of recycled fibers in fashion and home products, Textiles and Clothing Sustainability, Springer, Singapore, 2017, pp. 55–77.
[8] http://www.indoramaindia.com/p_staple_fibre.php.
[9] http://nptel.ac.in/courses/116102010/6.
[10] http://www.tikp.co.uk/knowledge/technology/fibre-and-filament-production/melt-spinning/.
[11] http://www.textilevaluechain.com/index.php/article/technical/item/247-recycling-of-plastic-bottles-into-yarn-fabric.
[12] https://oecotextiles.wordpress.com/2010/03/31/issues-with-using-recycled-polyester/.
[13] http://www.petrecycling.in/uses-of-recycled-pet/.
[14] http://www.komalfibres.com/komal-fibers/contact-komal-fibers.htm.
[15] https://frajorden.wordpress.com/2016/09/21/how-pet-bottles-convert-to-recycled-polyester-garments/.
[16] Thesis entitled "From PET bottles to Clothing: Does it have a future in fashion?" Submitted by Marion Weinans to Amsterdam Fashion Institute under the guidance of Jan de Vries, 2010.
[17] D. Paszun, T. Spychaj, Chemical recycling of poly (ethylene terephthalate), Ind. Eng. Chem. Res. 36 (4) (1997) 1373–1383.
[18] A. Al-Sabagh, F. Yehia, G. Eshaq, A. Rabie, A. ElMetwally, Greener routes for recycling of polyethylene terephthalate, Egypt. J. Petrol. 25 (1) (2016) 53–64.
[19] S. Akçaözoğlu, C.D. Atiş, K. Akçaözoğlu, An investigation on the use of shredded waste PET bottles as aggregate in lightweight concrete, Waste Manag. 30 (2) (2010) 285–290.
[20] S. Akçaözoğlu, C.D. Atiş, Effect of granulated blast furnace slag and fly ash addition on the strength properties of lightweight mortars containing waste PET aggregates, Constr. Build. Mater. 25 (10) (2011) 4052–4058.
[21] M. Frigione, Recycling of PET bottles as fine aggregate in concrete, Waste Manag. 30 (6) (2010) 1101–1106.
[22] A. Hassani, H. Ganjidoust, A.A. Maghanaki, Use of plastic waste (polyethylene terephthalate) in asphalt concrete mixture as aggregate replacement, Waste Manag. Res. 23 (4) (2005) 322–327.
[23] J.D. Rusmirović, A.R. Božić, M. Stamenović, P. Spasojević, M. Rančić, et al., Alkyd nanocomposite coatings based on waste pet glicolyzates and modified silica nanoparticles, Zaštita Materijala 57 (1) (2016) 47–54.
[24] I.L. Pogrebnyak, N.V. Sova, B.M. Savchenko, V.A. Pakharenko, V.S. Moisyuk, A wood-filled composite based on recycled polyethylene

terephthalate. Production and properties, Int. Polym. Sci. Technol. 42 (1) (2015) T41–T43.
[25] I. Ahmad, D.A. Bakar, S.N. Mokhilas, A. Ramli, Recycled PET for rice husk/polyester composites, ASEAN J. Sci. Technol. Dev. 22 (4) (2017) 345–353.
[26] S.H. Park, S.H. Kim, Poly (ethylene terephthalate) recycling for high value added textiles, Fashion Text. 1 (1) (2014) 1.
[27] M.E. Tawfik, S.B. Eskander, Polymer concrete from marble wastes and recycled poly (ethylene terephthalate), J. Elastom. Plast. 38 (1) (2006) 65–79.
[28] G.P. Karayannidis, D.S. Achilias, Chemical recycling of poly (ethylene terephthalate), Macromol. Mater. Eng. 292 (2) (2007) 128–146.
[29] S.F. Pusztaszeri, 1982. US Pat. 4, 355, 175.
[30] J. Datta, P. Kopczyńska, From polymer waste to potential main industrial products: Actual state of recycling and recovering, Crit. Rev. Environ. Sci. Technol. 46 (10) (2016) 905–946.
[31] M.E. Tawfik, N.M. Ahmed, S.B. Eskander, Aminolysis of poly (ethylene terephthalate) wastes based on sunlight and utilization of the end product [bis (2-hydroxyethylene) terephthalamide] as an ingredient in the anticorrosive paints for the protection of steel structures, J. Appl. Polym. Sci. 120 (5) (2011) 2842–2855.
[32] T. Spychaj, E. Fabrycy, S. Spychaj, M. Kacperski, Aminolysis and aminoglycolysis of waste poly (ethylene terephthalate), J. Mater. Cycles Waste Manag. 3 (1) (2001) 24–31.
[33] L.-N. Zhang, L.-Z. Liu, Q.-F. Yue, C.-C. Zhu, From aminolysis product of PET waste to value-added products of polymer and assistants, Polym. Polym. Compos. 22 (1) (2014) 13.
[34] W.M.N.W.A. Rahman, A.F.A. Wahab, Green pavement using recycled polyethylene terephthalate (PET) as partial fine aggregate replacement in modified asphalt, Proc. Eng. 53 (2013) 124–128.
[35] M.A. Kamall, Recycling of Waste PET Bottles as a Voluminous Alternative Material for Making Composite Panels.
[36] K. Nováková, K. Šeps, H. Achten, Experimental development of a plastic bottle usable as a construction building block created out of polyethylene terephthalate: Testing PET (b) rick 1.0, J. Build. Eng. 12 (2017) 239–247.
[37] F. Mahdi, H. Abbas, A.A. Khan, Flexural, shear and bond strength of polymer concrete utilizing recycled resin obtained from post consumer PET bottles, Constr. Build. Mater. 44 (2013) 798–811.
[38] M. Sogancioglu, A. Yucel, E. Yel, G. Ahmetli, Production of epoxy composite from the pyrolysis char of washed PET wastes, Energy Proc. 118 (2017) 216–220.
[39] N.I. Bento, P.S. Santos, T.E. de Souza, L.C. Oliveira, C.S. Castro, Composites based on PET and red mud residues as catalyst for organic removal from water, J. Hazard. Mater. 314 (2016) 304–311.

[40] M.E. Çınar, F. Kar, Characterization of composite produced from waste PET and marble dust, Constr. Build. Mater. 163 (2018) 734–741.

[41] J. Dullius, C. Ruecker, V. Oliveira, R. Ligabue, S. Einloft, Chemical recycling of post-consumer PET: Alkyd resins synthesis, Prog. Org. Coat. 57 (2) (2006) 123–127.

[42] G. Güçlü, Alkyd resins based on waste PET for water-reducible coating applications, Polym. Bull. 64 (2010) 739–748.

[43] K. Bal, K.C. Ünlü, I. Acar, G. Güçlü, Epoxy-based paints from glycolysis products of postconsumer PET bottles: synthesis, wet paint properties and film properties, J. Coat. Technol. Res. 14 (3) (2017) 747–753.

[44] C. Kawamura, K. Ito, R. Nishida, I. Yoshihara, N. Numa, Coating resins synthesized from recycled PET, Prog. Org. Coat. 45 (2–3) (2002) 185–191.

[45] S.M. Cakić, I.S. Ristić, M. Milena, N.Č. Nikolić, O.Z. Ilić, D.T. Stojiljković, et al., Glycolyzed products from PET waste and their application in synthesis of polyurethane dispersions, Prog. Org. Coat. 74 (1) (2012) 115–124.

[46] M.R. Patel, J.V. Patel, V.K. Sinha, Polymeric precursors from PET waste and their application in polyurethane coatings, Polym. Degrad. Stab. 90 (2005) 111–115.

[47] N.E. Ikladious, J.N. Asaad, H.S. Emira, S.H. Mansour, Alkyd resins based on hyperbranched polyesters and PET waste for coating applications, Prog. Org. Coat. 102 (2017) 217–224.

[48] M. Kathalewar, N. Dhopatkar, B. Pacharane, A. Sabnis, P. Raut, et al., Chemical recycling of PET using neopentyl glycol: Reaction kinetics and preparation of polyurethane coating, Prog. Org. Coat. 76 (1) (2013) 147–156.

[49] V. Jamdar, M. Kathalewar, R.N. Jagtap, K.A. Dubey, A. Sabnis, Effect of γ-irradiation on glycolysis of PET waste and preparation of eco-friendly coatings using bio-based and recycled materials, Polym. Eng. Sci. 55 (11) (2015) 2653–2660.

[50] A. Torlakoğlu, G. Güçlü, Alkyd–amino resins based on waste PET for coating applications, Waste Manag. 29 (1) (2009) 350–354.

[51] M. Evstatiev, S. Fakirov, B. Krasteva, K. Friedrich, J.A. Covas, A.M. Cunha, Recycling of poly (ethylene terephthalate) as polymer–polymer composites, Polym. Eng. Sci. 42 (4) (2002) 826–835.

[52] I. Pesneau, A.A. Kadi, M. Bousmina, P.H. Cassagnau, A. Michel, From polymer blends to in situ polymer/polymer composites: morphology control and mechanical properties, Polym. Eng. Sci. 42 (10) (2002) 1990–2004.

[53] Z.M. Li, M.B. Yang, B.H. Xie, J.M. Feng, R. Huang, In-situ microfiber reinforced composite based on PET and PE via slit die extrusion and hot stretching: influences of hot stretching ratio on morphology and tensile properties at a fixed composition, Polym. Eng. Sci. 43 (3) (2003) 615–628.

[54] Z.Z. Denchev, N.V. Dencheva, Transforming polymer blends into composites: a pathway towards nanostructured materials, Polym. Int. 57 (1) (2008) 11–22.
[55] K. Jayanarayanan, S. Thomas, K. Joseph, Morphology, static and dynamic mechanical properties of in situ microfibrillar composites based on polypropylene/poly (ethylene terephthalate) blends. Compos, Part A Appl. Sci. Manufact 39 (2) (2008) 164–175.
[56] M. Evstatiev, S. Fakirov, Microfibrillar reinforcement of polymer blends, Polymer. (Guildf). 33 (4) (1992) 877–880.
[57] Z.M. Li, W. Yang, L.B. Li, B.H. Xie, R. Huang, M.B. Yang, Morphology and nonisothermal crystallization of in situ microfibrillar poly (ethylene terephthalate)/polypropylene blend fabricated through slit-extrusion, hot-stretch quenching, J. Polym. Sci. Part B Polym. Phys. 42 (3) (2004) 374–385.
[58] Z.M. Li, M.B. Yang, B.H. Xie, J.M. Feng, R. Huang, In-situ microfiber reinforced composite based on PET and PE via slit die extrusion and hot stretching: influences of hot stretching ratio on morphology and tensile properties at a fixed composition, Polym. Eng. Sci. 43 (3) (2003) 615–628.
[59] X.B. Xu, Z.M. Li, M.B. Yang, S. Jiang, R. Huang, The role of the surface microstructure of the microfibrils in an electrically conductive microfibrillar carbon black/poly (ethylene terephthalate)/polyethylene composite, Carbon. N. Y. 43 (7) (2005) 1479–1487.
[60] S. Fakirov, D. Bhattacharyya, R.J.T. Lin, C. Fuchs, K. Friedrich, Contribution of coalescence to microfibril formation in polymer blends during cold drawing, J. Macromol. Sci. Part B Phys. 46 (1) (2007) 183–194.

Index

Note: Page numbers followed by "*f*" and "*t*" refer to figures and tables, respectively.

FSC
www.fsc.org
MIX
Papier | Fördert
gute Waldnutzung
FSC® C083411

Kolibri 360 GmbH
Ferdinand-Jühlke-Straße 7
99095 Erfurt, Deutschland
produktsicherheit@kolibri360.de